Dimensions Math
Tests 1A

Author
Dawn Yuen

Singapore Math Inc.

Published by Singapore Math Inc.

19535 SW 129th Avenue
Tualatin, OR 97062
www.singaporemath.com

Dimensions Math® Tests 1A
ISBN 978-1-947226-47-0

First published 2019
Reprinted 2020 (twice), 2021 (twice), 2022, 2023 (twice)

Copyright © 2018 by Singapore Math Inc.
All rights reserved. This book or any portion thereof may not be reproduced or used in any manner whatsoever without the express written permission of the publisher.

Printed in China

Acknowledgments

Design and illustration by Cameron Wray with Carli Bartlett.

Preface

Dimensions Math® Tests is a series of assessments to help teachers systematically evaluate student progress. The tests align with the content of Dimensions Math K–5 textbooks.

Dimensions Math Tests K uses pictorially engaging questions to test student ability to grasp key concepts through various methods including circling, matching, coloring, drawing, and writing numbers.

Dimensions Math Tests 1–5 have differentiated assessments. Tests consist of multiple-choice questions that assess comprehension of key concepts, and free response questions for students to demonstrate their problem-solving skills.

Test A focuses on key concepts and fundamental problem-solving skills.

Test B focuses on the application of analytical skills, thinking skills, and heuristics.

Contents

Chapter	Test	Page
Chapter 1 **Numbers to 10**	Test A Test B	1 7
Chapter 2 **Number Bonds**	Test A Test B	13 19
Chapter 3 **Addition**	Test A Test B	25 31
Chapter 4 **Subtraction**	Test A Test B Continual Assessment 1 Test A Continual Assessment 1 Test B	37 43 49 59
Chapter 5 **Numbers to 20**	Test A Test B	69 75
Chapter 6 **Addition to 20**	Test A Test B	81 87

Chapter	Test	Page
Chapter 7 **Subtraction Within 20**	Test A Test B	93 99
Chapter 8 **Shapes**	Test A Test B	105 111
Chapter 9 **Ordinal Numbers**	Test A Test B **Continual Assessment 2 Test A** **Continual Assessment 2 Test B**	117 123 129 139
Answer Key		149

BLANK

Name: _____

Date: _____

 25 min   Score

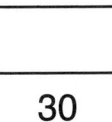 30

Test A

Chapter 1 Numbers to 10

Section A (2 points each)
Circle the correct option: **A**, **B**, **C**, or **D**.

1 How many dots are there?

•	•	•		
•	•			

A 7   **B** 6

C 5   **D** 4

2 Which bowl has 0 strawberries?

A    B

C    D

Chapter 1 Test A   1

3) What is the missing number?

| 9 | 8 | | 6 | 5 | 4 |

A 10

B 7

C 3

D 6

4) Which box is missing?

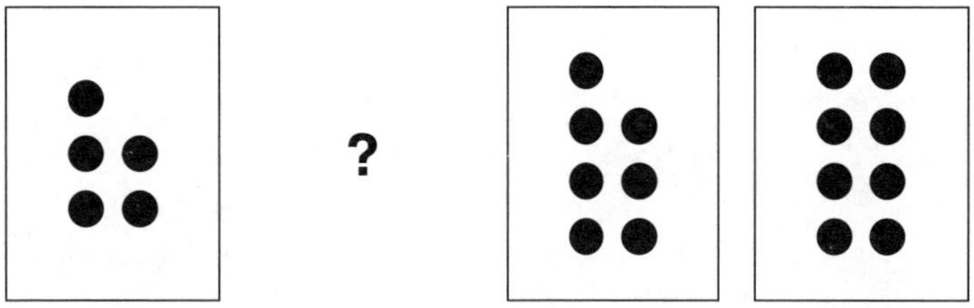

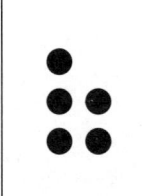   ?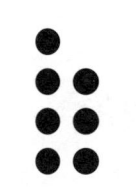

A

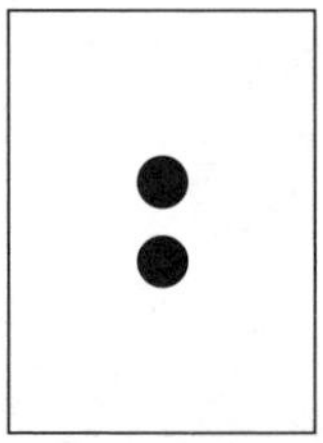

B

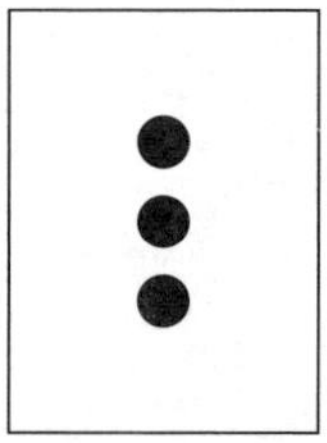

C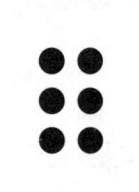

D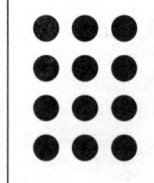

5) Which number is greatest?

A 6

B 7

C 3

D 4

Section B (2 points each)

6 Match.

7 How many bees are there?
Check (✓) the correct answer.

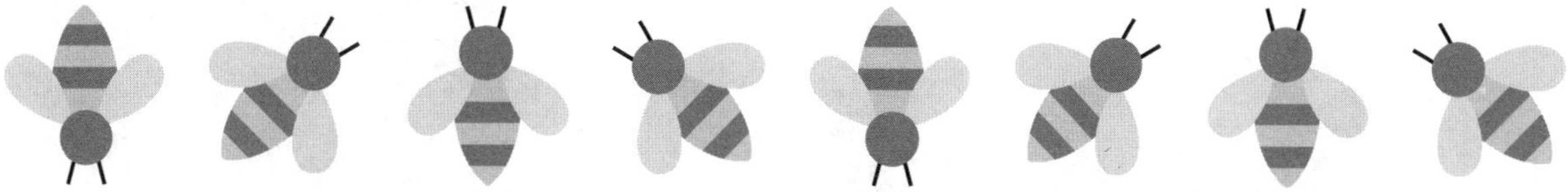

two	zero	eight	nine
☐	☐	☐	☐

8 Circle the bowl with 0 cherries.

Chapter 1 Test A   3

9 Write the numbers.

three ☐

five ☐

10 Circle the number that is greater than 6.

| 7 | 5 | 0 | 4 |

11 Cross out the set that has fewer.

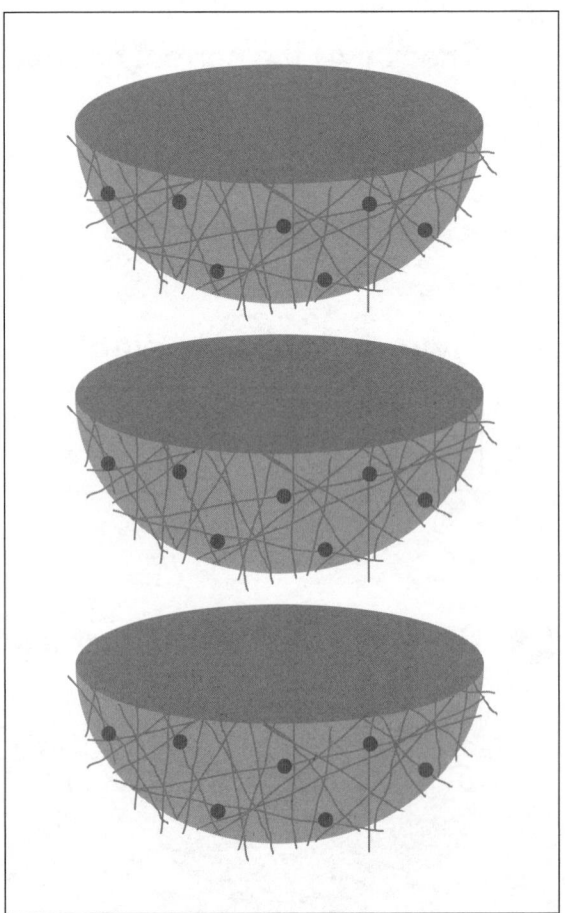

Chapter 1 Test A

12 Check (✓) the set that has 1 more than 5 ants.

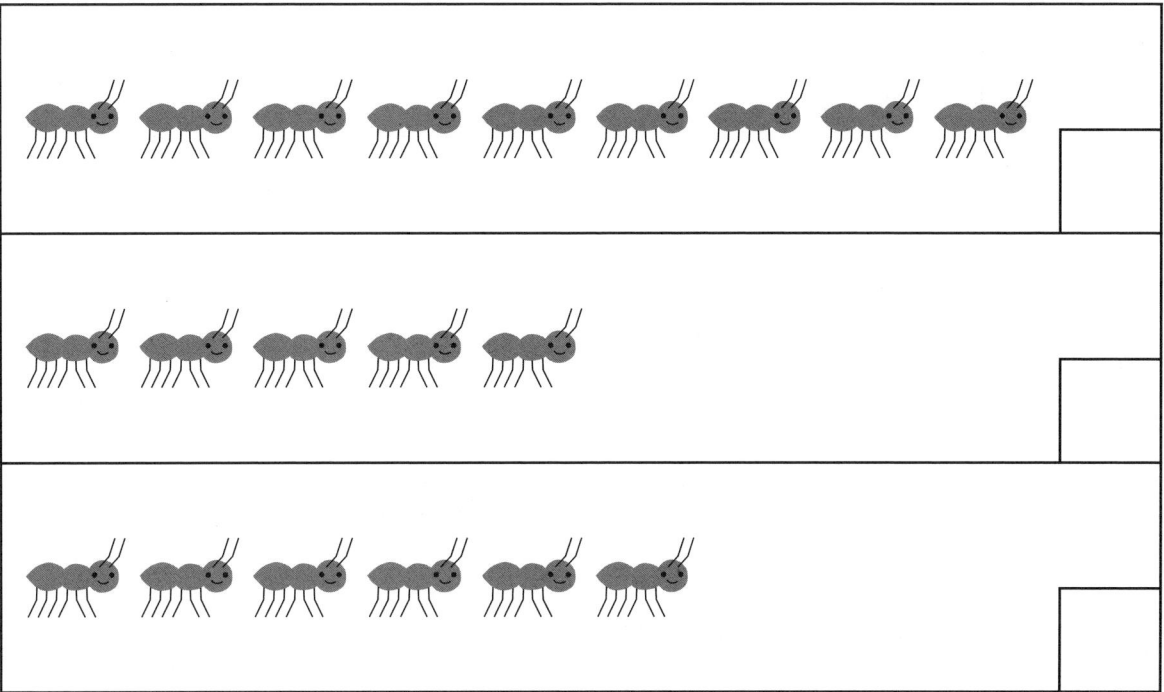

13 Which number is least?
Circle it.

| three   five   ten   six |

Chapter 1 Test A

14 Write the number that is 1 less than 10.

☐

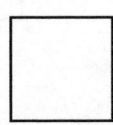

15 Write the missing numbers.

| 2 | 3 | | 5 | 6 | |

Name: _____

Date: _____

25 min

Score

30

Test B

Chapter 1 Numbers to 10

Section A (2 points each)

Circle the correct option: **A**, **B**, **C**, or **D**.

1 Which set does not have 8?

A

B

C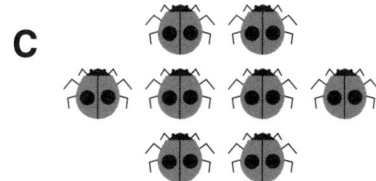

D

2 Which number goes in the white box?
Some numbers are hidden.

A 7

B 8

C 9

D 5

Chapter 1 Test B

7

3 What number is 1 more than 9?

    A 4   **B** 8

    C 7   **D** 10

4 Which number is greater than 3 and less than 6?

    A 4   **B** 8

    C 0   **D** 7

5 Which box is missing?

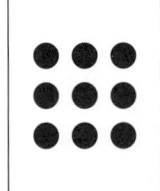 ?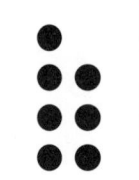

A 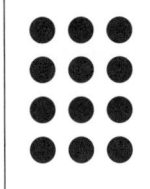   **B**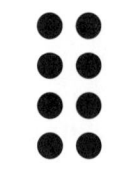

C 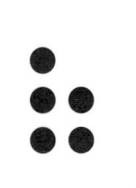   **D**

Section B (2 points each)

6 Write the numbers.

zero ☐

ten ☐

7 Write the missing number word.

four   five   _____   seven   eight

8 Check (✓) the set that has 1 less than 10.

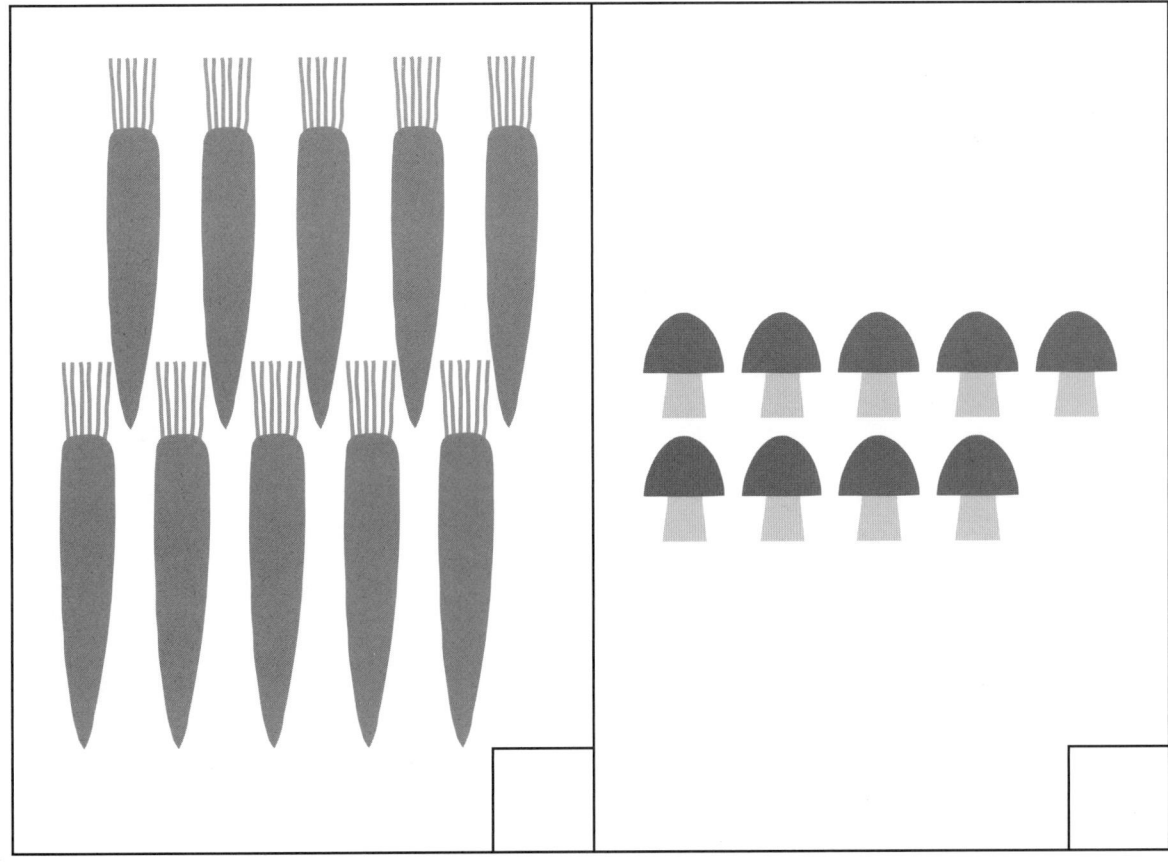

9 Match.

10 Write the missing numbers.

| 8 | 7 | | 5 | | 3 | 2 |

11 Ivan eats more than 6 but less than 8 strawberries.
Circle the number of strawberries that he eats.

12 Cross out the greatest number.

| 7 | 5 | 0 | 9 | 8 |

Use the numbers below to answer questions 13 to 15.

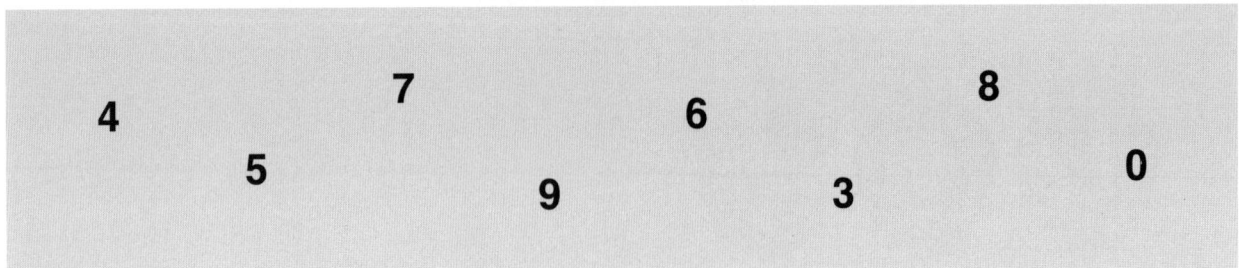

13 What number is 9 more than 0?

_____

14 What number is 2 less than 2?

_____

15 What numbers are greater than 4 and less than 8?

_____

12 Chapter 1 Test B

Name: _____

Date: _____

25 min **Score**

30

Test A

Chapter 2 Number Bonds

Section A (2 points each)
Circle the correct option: **A**, **B**, **C**, or **D**.

1 How many more ● make 7?

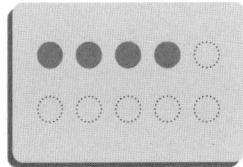

A 2   **B** 3

C 4   **D** 7

2 What is the missing number?

A 1   **B** 5

C 6   **D** 2

Chapter 2 Test A   13

3 _____ and 2 make 9.

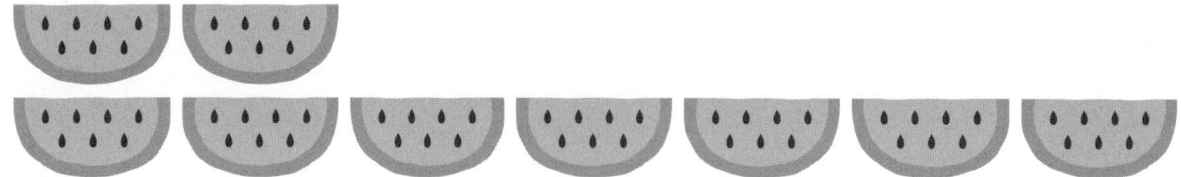

A 9

B 7

C 2

D 0

4 What is the missing number?

 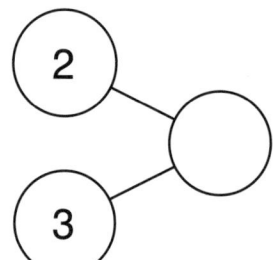

A 4

B 6

C 5

D 3

5 What is the missing number?

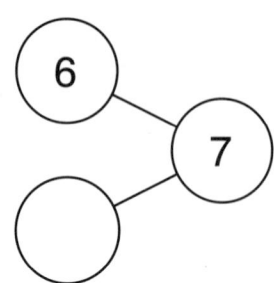

A 1

B 2

C 3

D 4

Section B (2 points each)

6 Write the missing number.

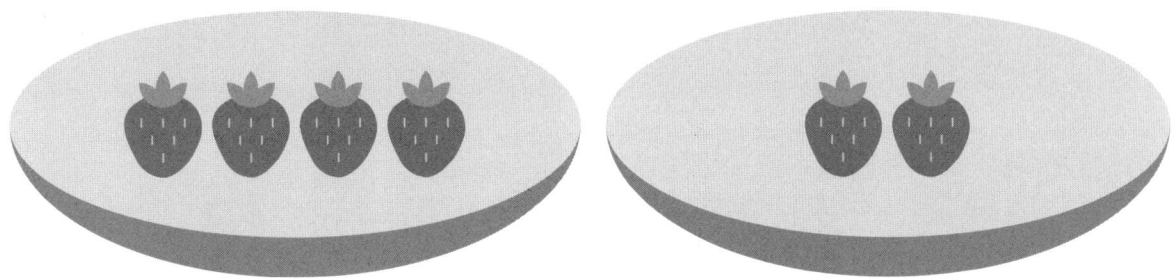

4 and 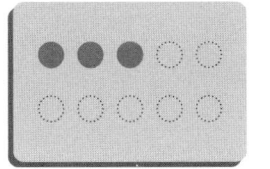 make 6.

7 Write the missing number.

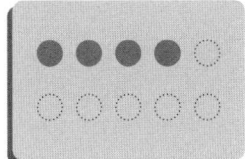

3 and [ ] make 7.

8 Write the missing number.

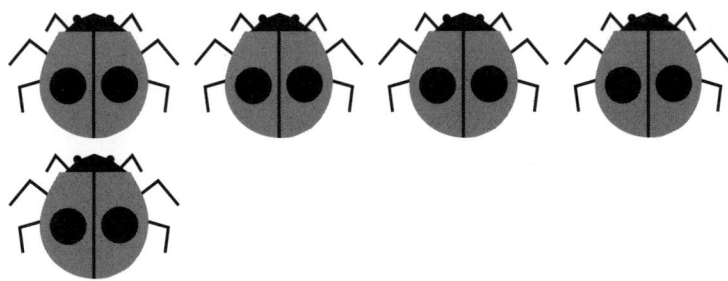

5 is 1 and [ ].

Chapter 2 Test A 15

9 Complete the number bond.

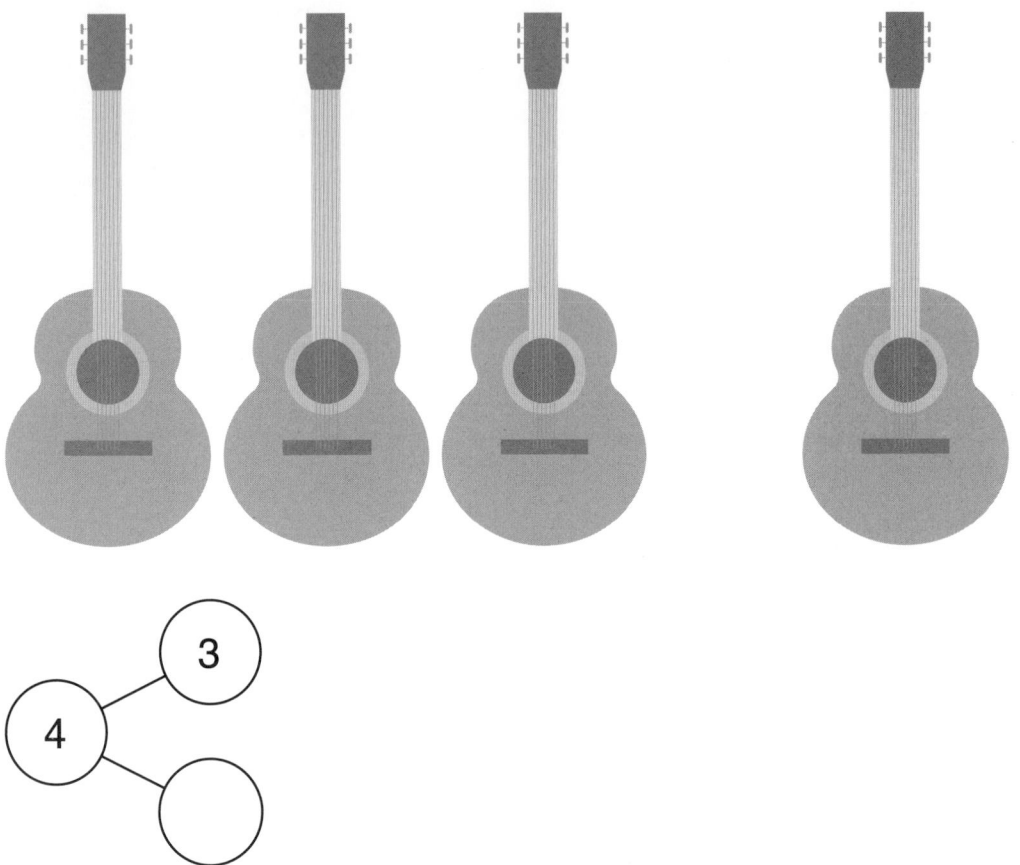

10 Complete the number bond.

11 Draw more apples to make 7, then write the missing number.

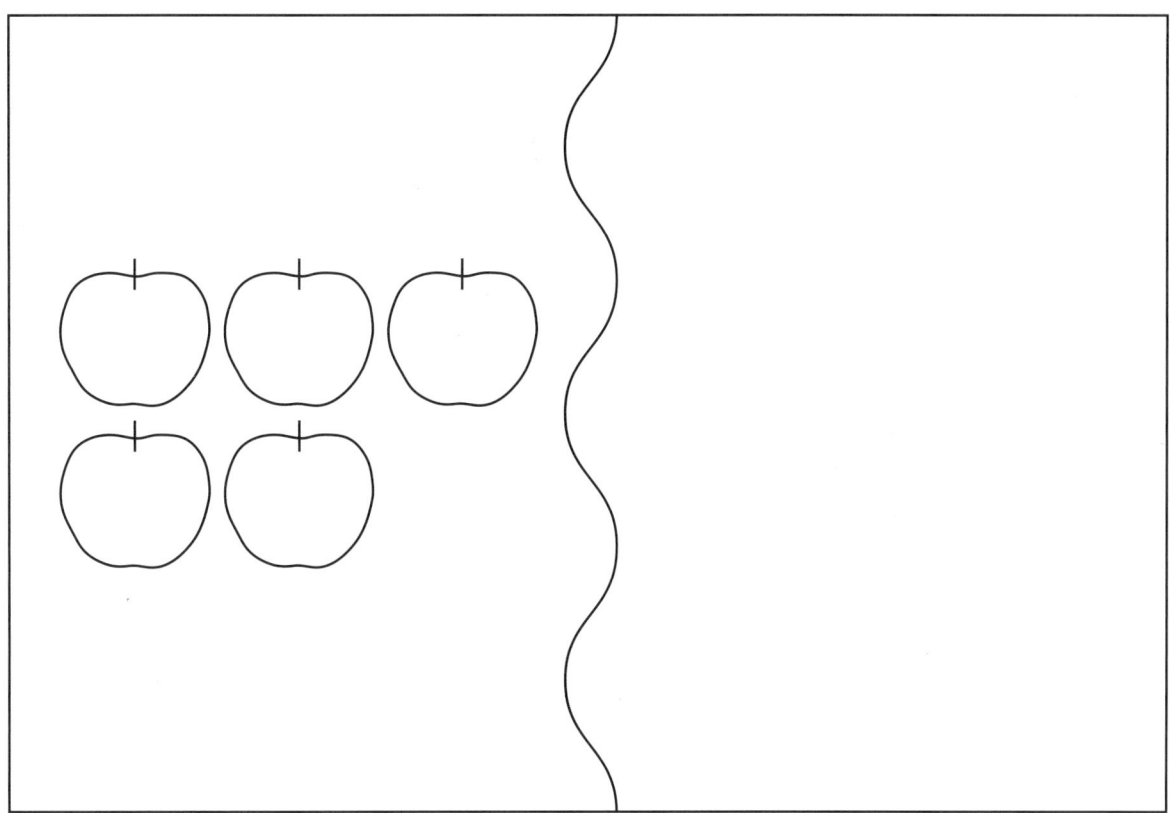

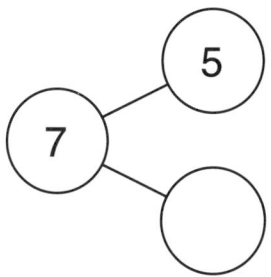

12 Write the missing numbers.

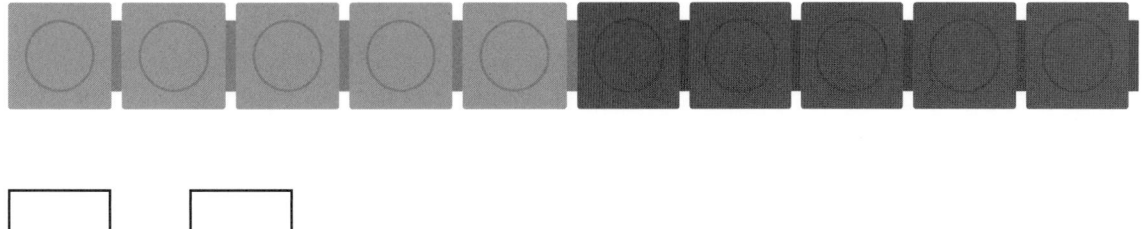

☐ and ☐ make 10.

13 Circle the two cards that together make 8.

| 6 | 1 | 3 | 2 | 4 |

14 Complete the number bond.

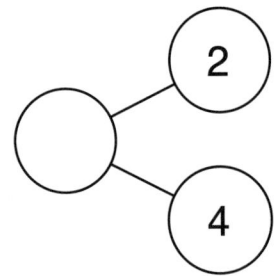

15 Complete the number bond.

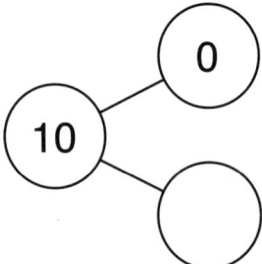

Name: _____

Date: _____

25 min   Score

30

Test B

Chapter 2 Number Bonds

Section A (2 points each)
Circle the correct option: **A**, **B**, **C**, or **D**.

1 How many more ● do we need to make 5?

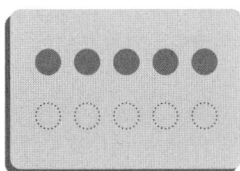

A 0   **B** 2

C 4   **D** 5

2 What is the missing number?

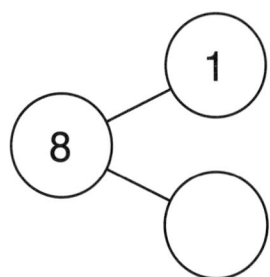

A 5   **B** 3

C 7   **D** 6

Chapter 2 Test B   19

3 What is the missing number?

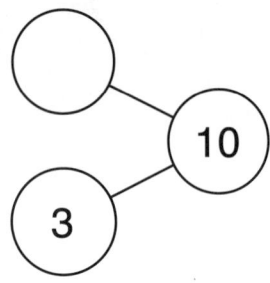

A 5

B 6

C 8

D 7

4 What is the missing number?

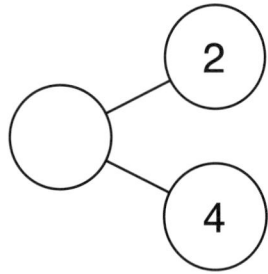

A 6

B 3

C 1

D 2

5 Which pair of numbers makes 8?

A 8 and 1

B 3 and 4

C 5 and 3

D 2 and 7

Section B (2 points each)

6 Write the missing number.

6 is ☐ and 2.

7 Write the missing number.

4 and 4 make ☐.

8 Color the 2 numbers that make 8.

6	4	3
1	5	6
1	9	8

Chapter 2 Test B

9 Circle the 2 numbers that make 10.

| 7 | 2 | 6 | 5 | 3 |

10 Complete the number bond.

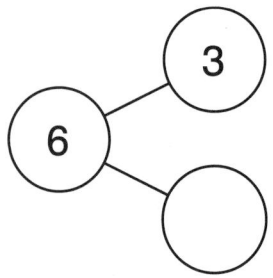

11 Complete the number bond.

12 Complete the number bond.

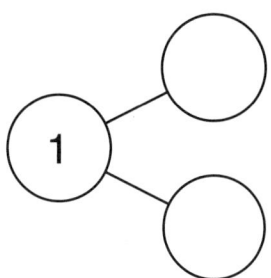

13 Complete the number bonds.

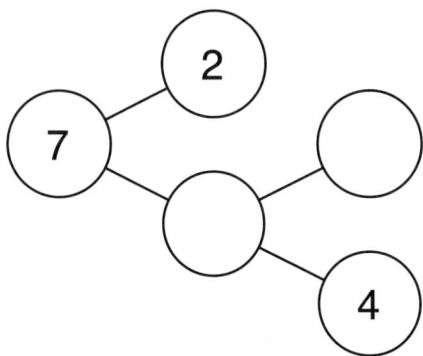

14 Complete the number bonds.

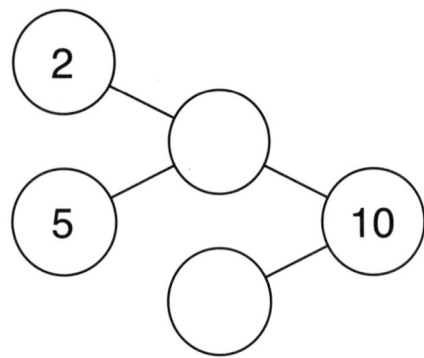

15 Use these numbers to make a number bond.

3, 6, 3

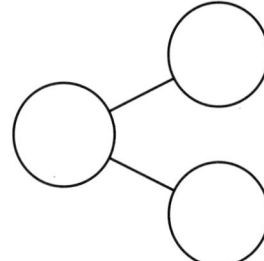

Name: _____

Date: _____

25 min   Score

30

Test A

Chapter 3 Addition

Section A (2 points each)
Circle the correct option: **A**, **B**, **C**, or **D**.

1 2 + 3 = _____

A 3   **B** 5

C 6   **D** 1

2 3 added to 4 is _____.

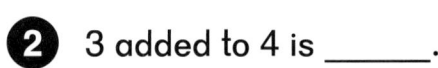

A 4   **B** 7

C 8   **D** 6

Chapter 3 Test A   25

3 What is 9 more than 0?

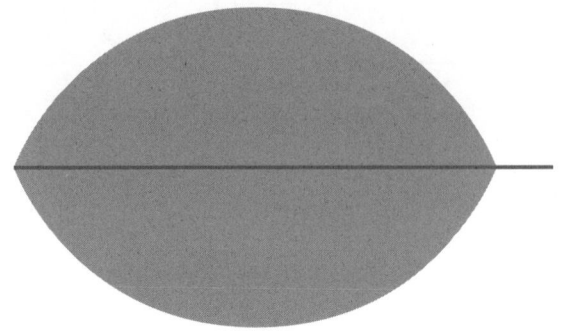 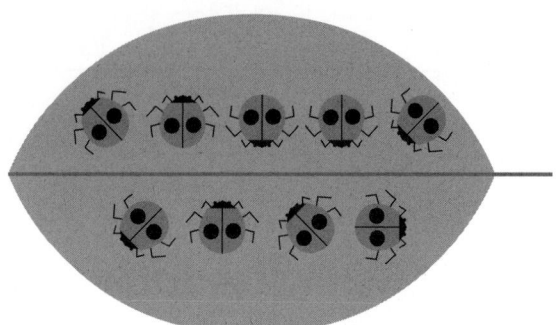

A 0

B 7

C 8

D 9

4 Which equation represents the number bond?

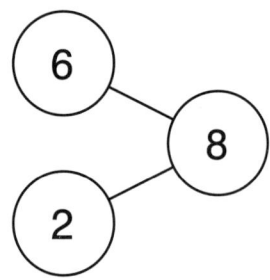

A 5 + 1 = 6

B 6 + 2 = 8

C 2 + 4 = 6

D 6 + 3 = 9

5 Which equation is true?

A 2 + 5 = 5

B 9 = 3 + 6

C 3 + 7 = 4

D 10 = 0 + 8

Section B (2 points each)

6 Write the missing number.

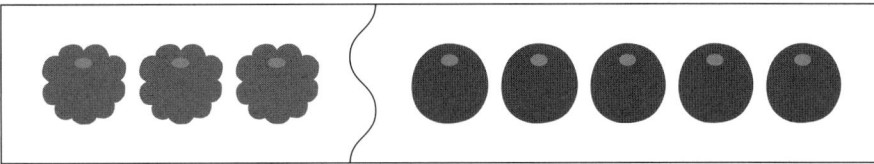

3 + 5 = ☐

7

☐ is 2 more than 4.

8 Write the missing numbers.

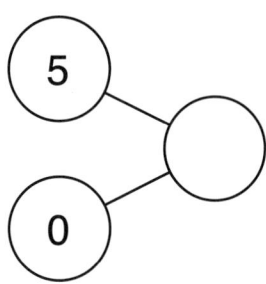

0 + ☐ = 5

9 Write the missing numbers.

6 + 3 = ☐

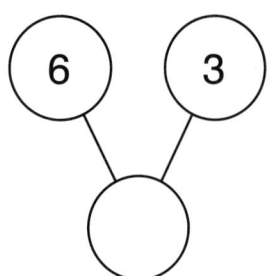

10 Complete the equation.

4 + ☐ = ☐

11 Write an equation.

☐ + ☐ = ☐

12 Write the missing numbers.

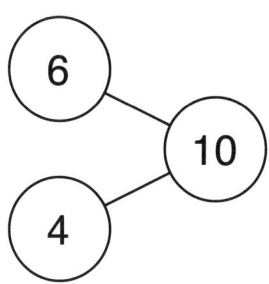

6 + ☐ = ☐

13 How many berries are there altogether?

 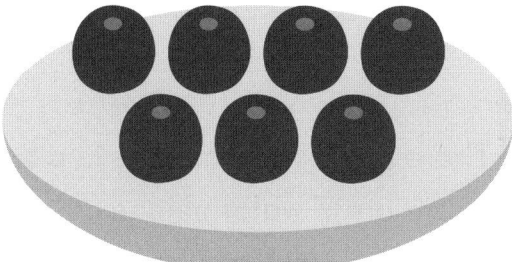

☐ + ☐ = ☐

There are _____ berries on the two plates.

Chapter 3 Test A

14 5 bats are on a branch.
1 more bat comes.
How many bats are there in all?

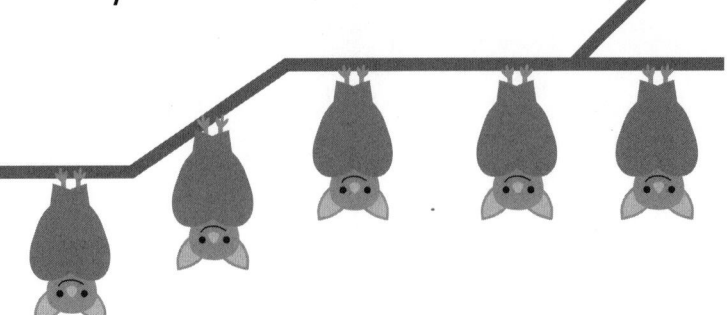

☐ + ☐ = ☐

There are _____ bats in all.

15 There are 2 watermelons in a basket.
There are 7 watermelons outside the basket.
How many watermelons are there altogether?

☐ ◯ ☐ = ☐

There are _____ watermelons altogether.

Name: _____

Date: _____

25 min   Score

30

Test B

Chapter 3 Addition

Section A (2 points each)

Circle the correct option: **A**, **B**, **C**, or **D**.

1 Which is the correct equation?

    A 2 + 7 = 9   **B** 5 + 5 = 10

    C 2 + 5 = 7   **D** 2 + 2 = 4

2 Which equation represents the number bond?

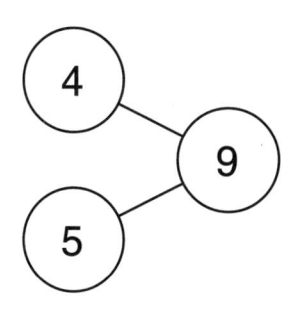

    A 4 + 1 = 5   **B** 5 + 9 = 4

    C 5 + 4 = 9   **D** 9 + 4 = 5

Chapter 3 Test B

3 0 added to 0 is _____.

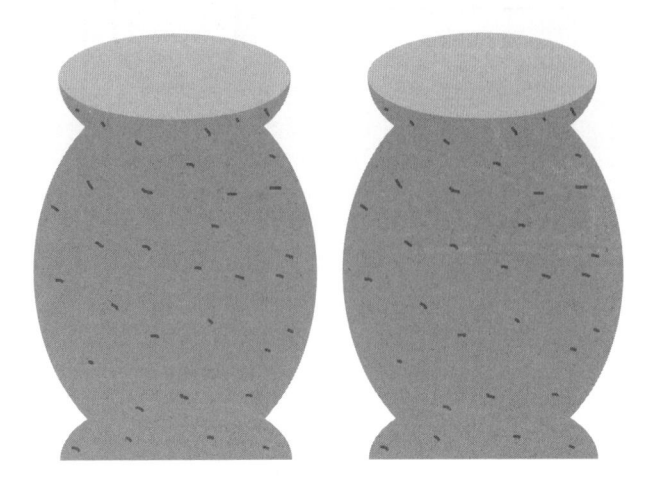

A 1

B 0

C 4

D 10

4 Which equation is false?

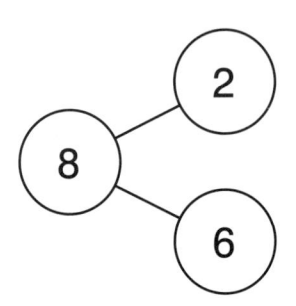

A 2 + 6 = 8

B 8 + 2 = 6

C 6 + 2 = 8

D 8 = 2 + 6

5 Which equation is true?

A 3 + 5 = 5 + 8

B 9 + 0 = 3 + 7

C 3 + 6 = 5 + 4

D 1 + 8 = 2 + 6

Section B (2 points each)

6 Write the missing numbers.

3 + ☐ = ☐

7 Complete the equation.

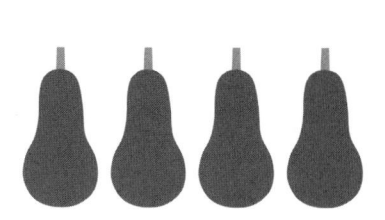

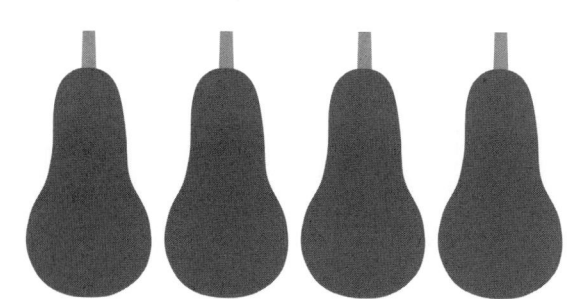

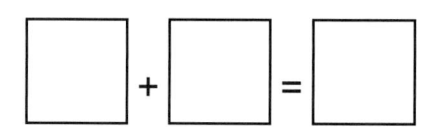

☐ + ☐ = ☐

Chapter 3 Test B

8 Write the missing numbers.

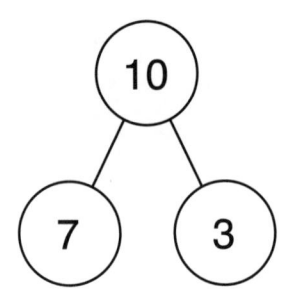

```
    10
   /  \
  7    3
```

7 + ☐ = ☐

9 Complete the equation.

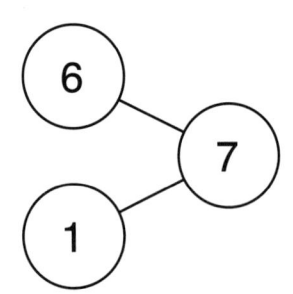

```
  6
   \
    7
   /
  1
```

☐ + 6 = ☐

10 2 more than 5 is ☐.

11 8 and 1 make ☐.

12 Complete the equation.

6 + 2 = ☐ + 7

13 Use the numbers and signs to write 2 equations.

10, 9, 1, =, +

_____

_____

14 There are 0 cookies in a jar.
Emma puts 3 cookies in the jar.
How many cookies are in the jar altogether?

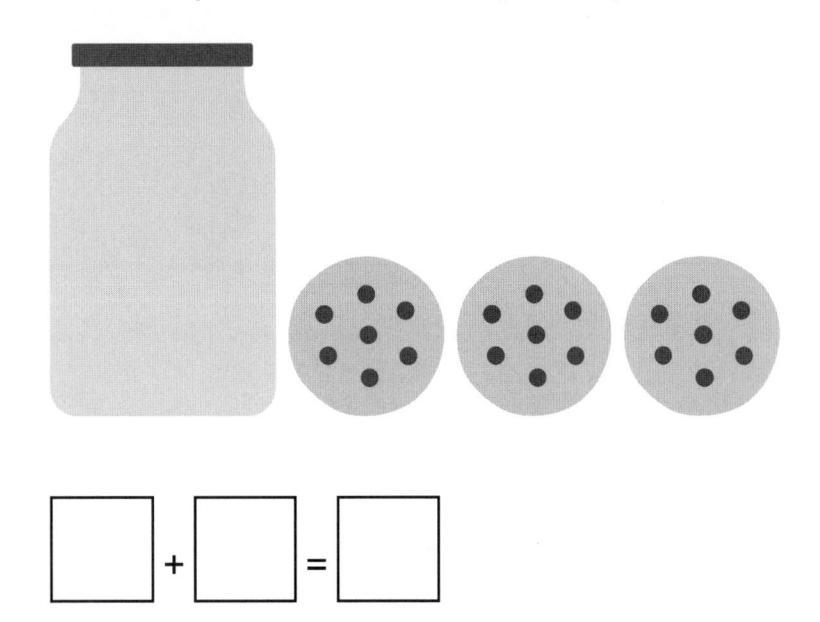

☐ + ☐ = ☐

There are _____ cookies in the jar altogether.

15 Dion has 6 baseball cards.
Alex gives him 4 more baseball cards.
How many baseball cards does Dion have in all?

☐ ◯ ☐ = ☐

Dion has _____ baseball cards in all.

Name: _____

Date: _____

25 min

Score

30

Test A

Chapter 4 Subtraction

Section A (2 points each)
Circle the correct option: **A**, **B**, **C**, or **D**.

1 6 − 3 = _____

A 6   **B** 3

C 0   **D** 1

2 What is 2 less than 8?

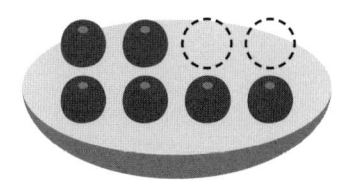

A 5   **B** 4

C 8   **D** 6

Chapter 4 Test A   37

3 Which equation tells you how many tops are spinning?

A 7 − 5

B 7 − 2

C 5 + 2

D 2 + 5

4 Which equation does not represent the picture?

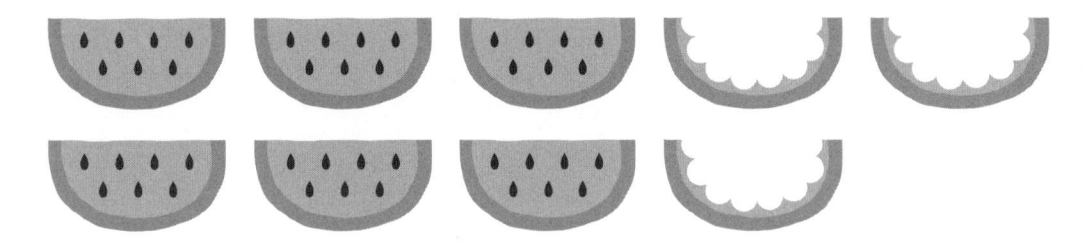

A 9 − 6 = 3

B 6 − 3 = 3

C 3 + 6 = 9

D 9 − 3 = 6

5 Which equation represents the number bond?

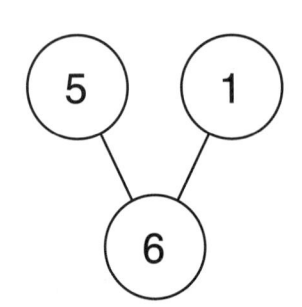

A 5 − 1 = 4

B 1 − 5 = 6

C 6 − 1 = 5

D 5 − 1 = 6

Section B (2 points each)

6 Write the missing number.

8 − 5 = ☐

7 Write the missing numbers.

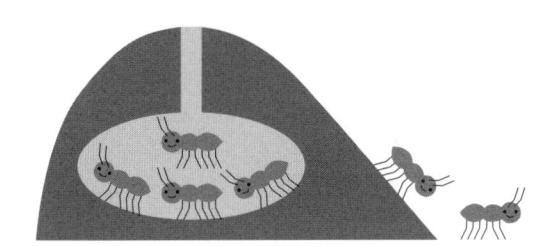

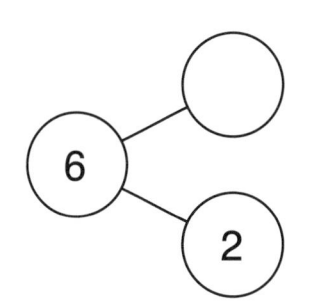

6 − ☐ = 4

8 Complete the equation.

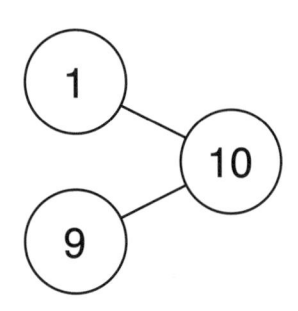

☐ − 9 = ☐

Chapter 4 Test A

9 Write two different subtraction equations.

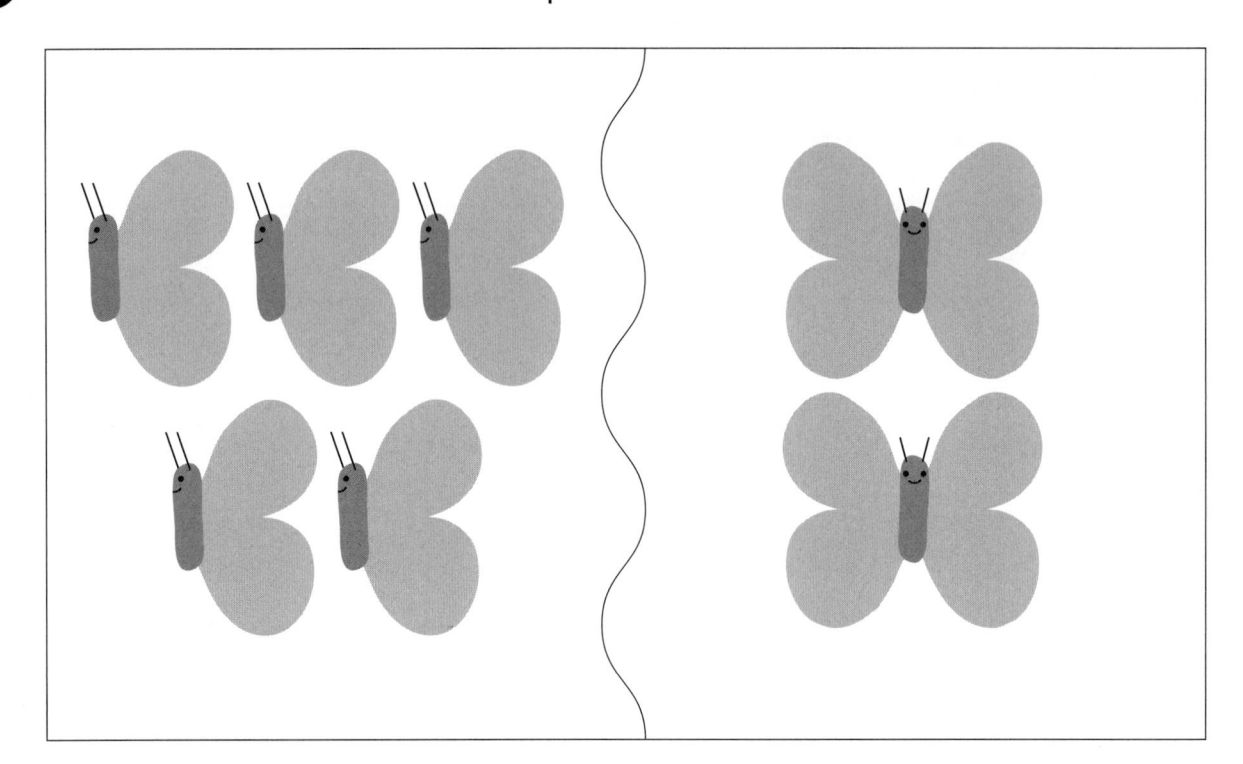

7 − ☐ = ☐

7 − ☐ = ☐

10 Write 2 different addition and 2 different subtraction equations.

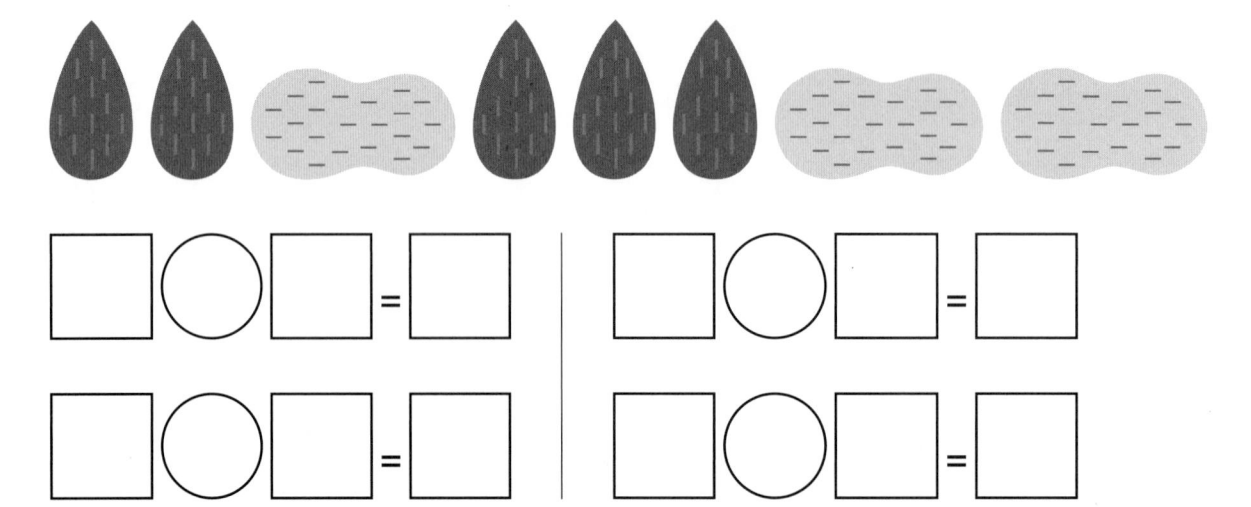

☐ ◯ ☐ = ☐   ☐ ◯ ☐ = ☐

☐ ◯ ☐ = ☐   ☐ ◯ ☐ = ☐

11 Complete an equation to find the number of candles still burning.

☐ − ☐ = ☐

12 There are 5 balloons.
5 balloons burst.
How many balloons are left?

5 − ☐ = ☐

_____ balloons are left.

13 Mei has 8 hair bands.
2 are in her pocket.
How many of her hair bands are not in her pocket?

☐ − ☐ = ☐

_____ of her hair bands are not in her pocket.

14 There are 9 children in a room.
5 of them are girls.
How many boys are in the room?

□ ○ □ = □

There are _____ boys in the room.

15 There are 10 children on a bus.
2 children get off the bus.
How many children are left on the bus?

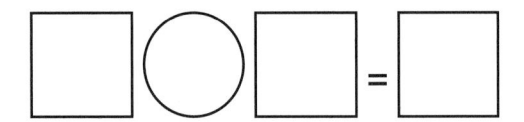

□ ○ □ = □

There are _____ children left on the bus.

Name: _____

Date: _____

25 min   Score

30

Test B

Chapter 4 Subtraction

Section A (2 points each)
Circle the correct option: **A**, **B**, **C**, or **D**.

1 Which one represents the picture?

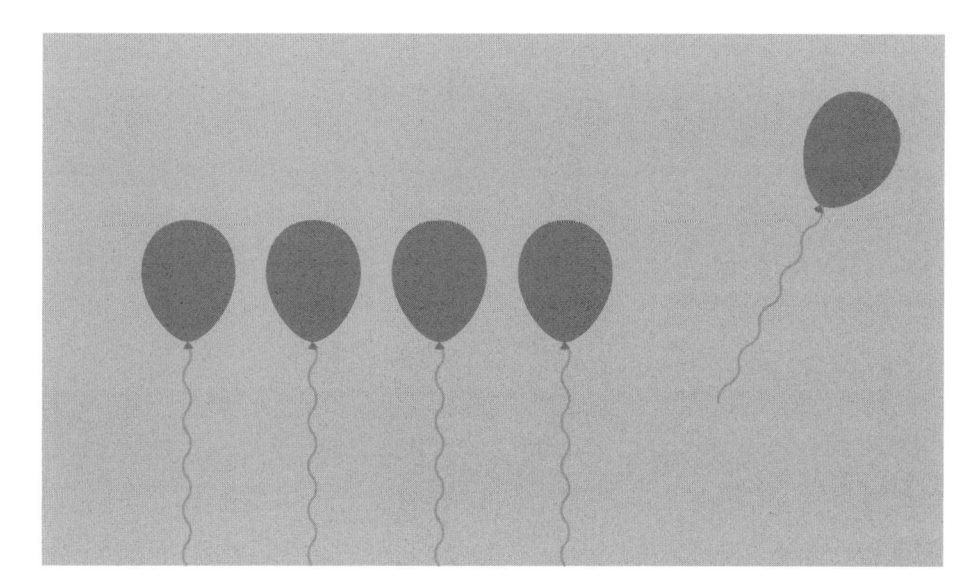

A 4 + 2   **B** 4 – 1

C 5 – 1   **D** 6 – 2

2 Which one does not give an answer of 2?

A 8 – 6   **B** 5 – 3

C 9 – 7   **D** 6 – 2

Chapter 4 Test B   43

3 2 less than 6 is the same as _____.

    A 1 less than 3   **B** 3 less than 8

    C 1 less than 5   **D** 2 less than 4

4 Which one gives the same answer as 10 − 3?

    A 7 − 0   **B** 8 − 2

    C 7 − 3   **D** 9 − 7

5 Which equation is false?

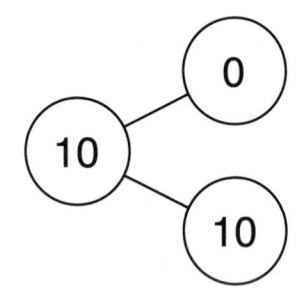

    A 10 − 10 = 0   **B** 0 + 10 = 10

    C 10 + 10 = 0   **D** 10 − 0 = 10

Section B (2 points each)

6 Write the missing number.

$9 - 7 =$ ☐

7 Write the missing numbers.

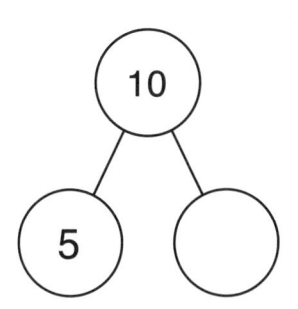

$10 - $ ☐ $=$ ☐

8 How many eggs have not hatched?

$8 - $ ☐ $=$ ☐

_____ eggs have not hatched.

Chapter 4 Test B 45

9 Write 2 different subtraction equations for the number bond.

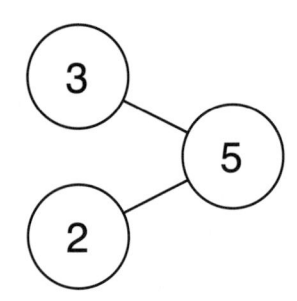

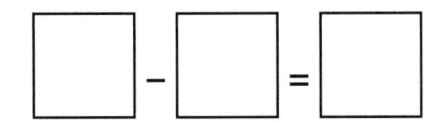

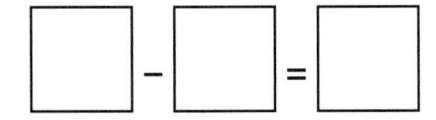

(3)
(5)
(2)

☐ − ☐ = ☐

☐ − ☐ = ☐

10 Write 1 addition and 1 subtraction equation for the number bond.

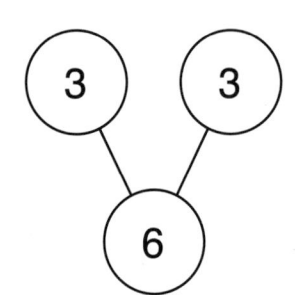

(3) (3)
(6)

☐ ◯ ☐ = ☐

☐ ◯ ☐ = ☐

11 Use these numbers to write 2 different addition and 2 different subtraction equations.

7, 10, 3

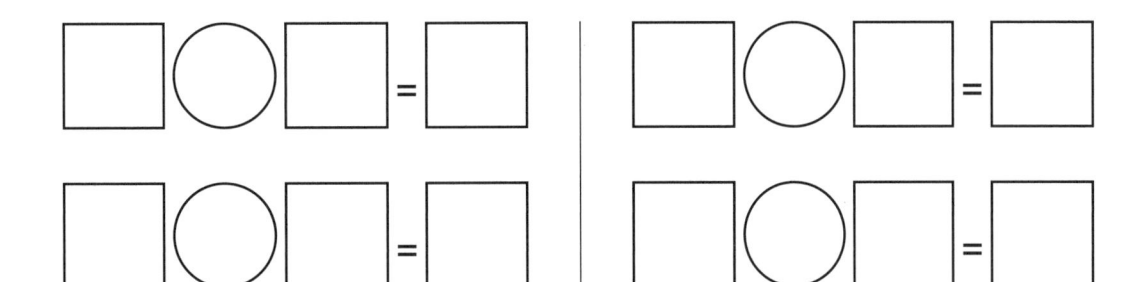

☐ ○ ☐ = ☐   ☐ ○ ☐ = ☐
☐ ○ ☐ = ☐   ☐ ○ ☐ = ☐

12 There are 8 children at the playground.
3 are wearing hats.
How many are not wearing hats?

☐ − 3 = ☐

There are _____ children not wearing hats.

Chapter 4 Test B

13 9 children are at a party.
5 children leave the party.
How many children are still at the party?

☐ − ☐ = ☐

_____ children are still at the party.

14 Dion and Sofia have 10 baseball cards altogether.
If Dion has 4 baseball cards, how many baseball cards does Sofia have?

☐ ○ ☐ = ☐

Sofia has _____ baseball cards.

15 7 birds are sitting on a fence.
Some fly away.
There are still 2 birds sitting on the fence.
How many birds fly away?

☐ ○ ☐ = ☐

_____ birds fly away.

Name: _____

Date: _____

45 min   Score

60

Test A

Continual Assessment 1

Section A (2 points each)
Circle the correct option: **A**, **B**, **C**, or **D**.

1 How many cups are there?

A 4

B 6

C 5

D 7

2 6 more than 3 is _____.

A 3

B 10

C 9

D 6

3 Which number is 1 more than 7?

A 8

B 6

C 7

D 1

4 What is the missing number?

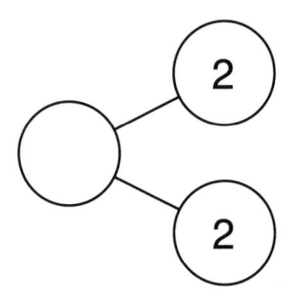

A 2

B 3

C 0

D 4

5 _____ and 7 make 9.

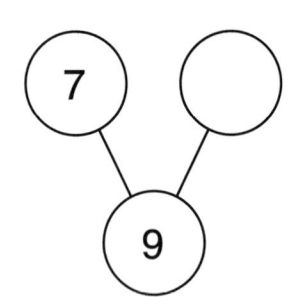

A 1

B 2

C 3

D 4

6 Which one represents the total number of fish?

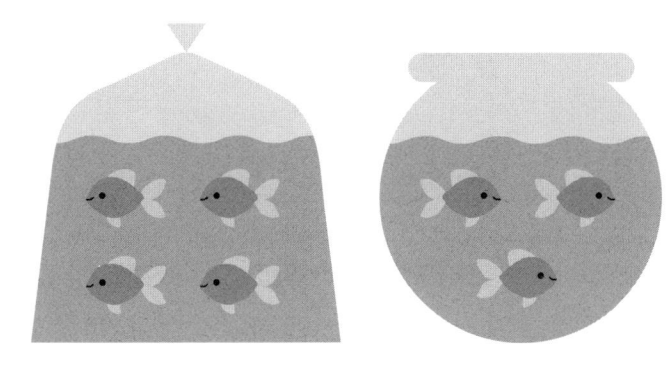

A 3 + 3

B 4 + 7

C 1 + 1

D 3 + 4

7 Which one represents the picture?

A 4 − 2

B 6 − 2

C 2 + 6

D 4 + 0

8 Which one has the answer 4?

A 6 − 3

B 9 − 4

C 8 − 5

D 4 − 0

Continual Assessment 1 Test A

9 Which equation is true?

  A 9 − 6 = 4

  B 3 + 5 = 7

  C 6 + 4 = 10

  D 7 − 7 = 1

10 There are 8 toys in the bucket.
There are 2 toys outside the bucket.
How many toys are there altogether?

  A 6

  B 10

  C 2

  D 9

Section B (2 points each)

11 How many hats are there?
Circle the answer.

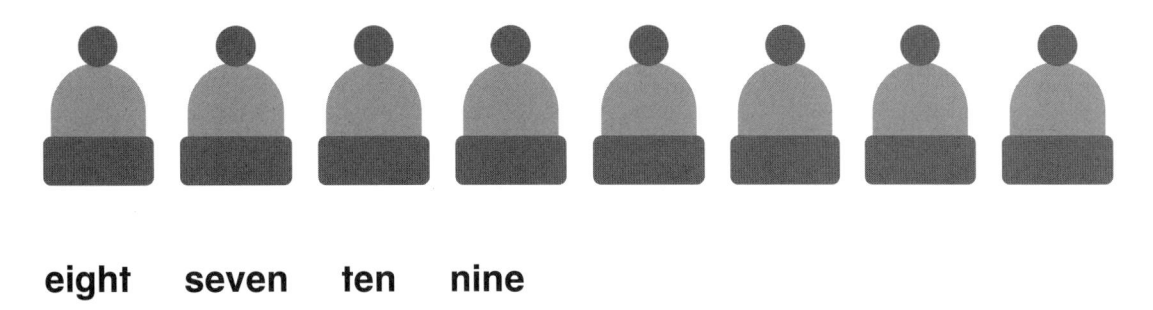

eight   seven   ten   nine

12 Cross out the group that has fewer than 7.

13 Circle the numbers that are greater than 8.

9   5   6   10   8

14 Write the missing numbers.

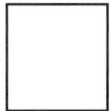

| 2 | | 4 | | 6 | |

15 Write the numbers in order from least to greatest.

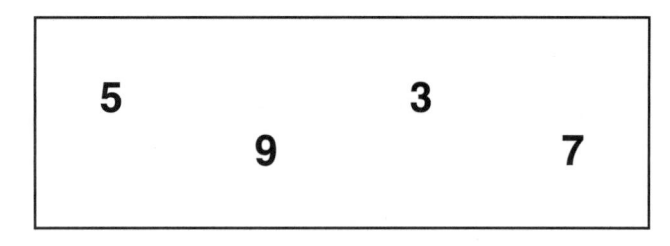

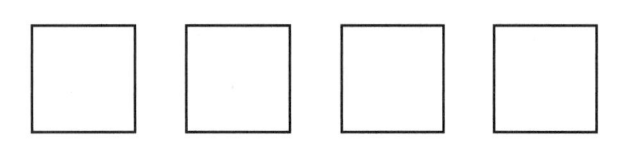

| 5   3 |
| 9   7 |

☐ ☐ ☐ ☐

16 Complete the number bond.

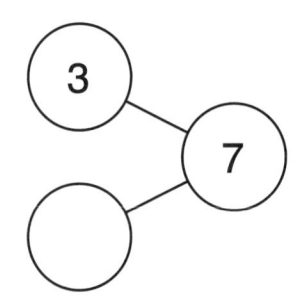

17 Complete the number bond.

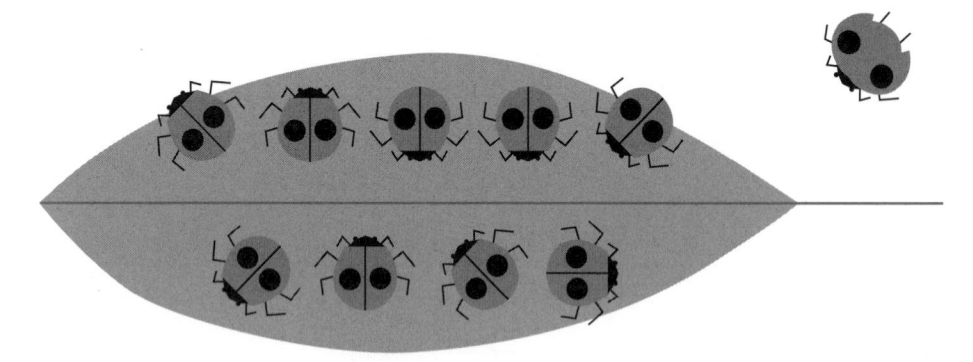

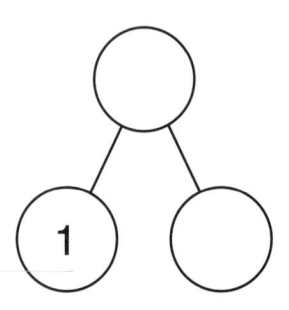

54   Continual Assessment 1 Test A

18 Complete the number bond with these numbers.

| 4 | 8 | 4 |

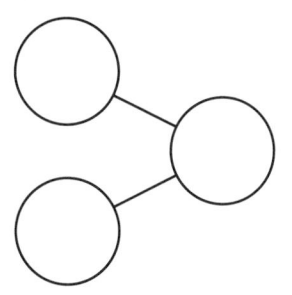

19 Complete the equation.

☐ + 3 = ☐

20 Complete the equation.

☐ − ☐ = 3

Section C (4 points each)

21 There are 7 limes in a bag.
There are 3 limes in a bowl.
How many limes are there in all?

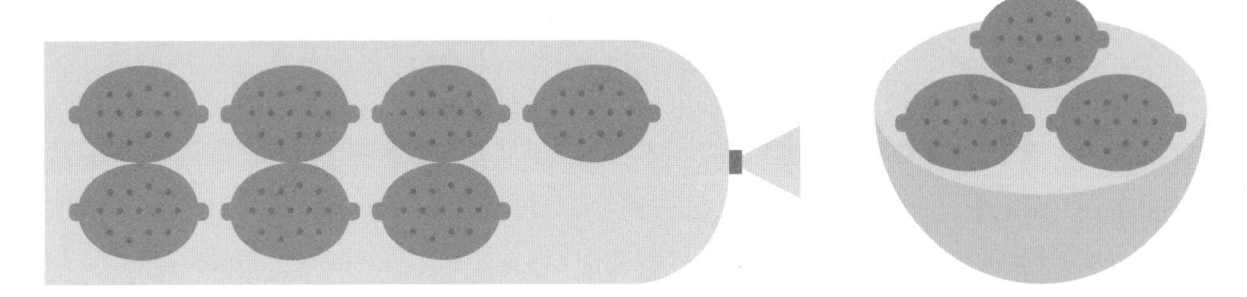

7 ◯ ▢ = ▢

There are _____ limes in all.

22 Chapa has 4 eggs.
She breaks 1 of the eggs.
How many eggs are not broken?

▢ ◯ ▢ = ▢

_____ eggs are not broken.

56 Continual Assessment 1 Test A

23 There are 9 apples.
6 apples are still on the tree.
How many apples fell off the tree?

☐ ◯ ☐ = ☐

_____ apples fell off the tree.

24 Emma has 4 stickers.
Dion gives her 6 more stickers.
How many stickers does she have?

☐ ◯ ☐ = ☐

Emma has _____ stickers.

25 There are 7 slices of pizza.
Sofia eats 2 slices of pizza.
How many slices of pizza are left?

☐ ◯ ☐ = ☐

There are _____ slices of pizza left.

Name: _____

Date: _____

45 min Score

60

Test B

Continual Assessment 1

Section A (2 points each)
Circle the correct option: **A**, **B**, **C**, or **D**.

1 Which set has 3 more than 0?

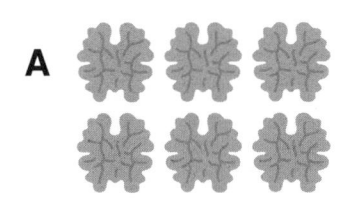

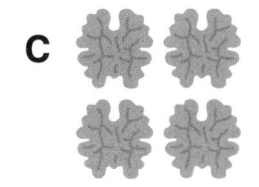

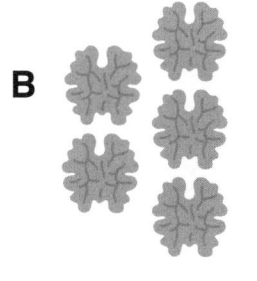

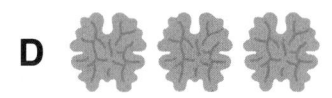

A   B

C   D

2 Take away one dot.
How many dots are left?

A 9   B 7

C 8   D 10

3 Which number is 9 less than 9?

    A seven   **B** zero

    C nine   **D** one

4 If one more mango is added to the box, there will be _____ mangoes.

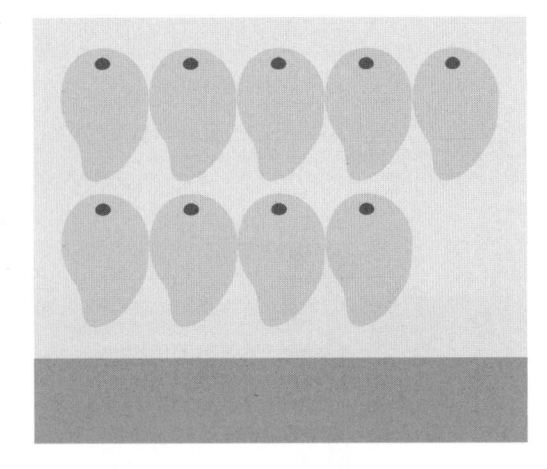

    A ten   **B** eight

    C nine   **D** one

5 What is the missing number?

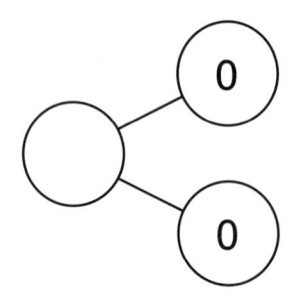

    A 1   **B** 3

    C 2   **D** 0

6 8 is _____ and 6.

　A 2　　　　　　　　　　**B** 6

　C 3　　　　　　　　　　**D** 5

7 4 + ☐ = 9

　A 6　　　　　　　　　　**B** 4

　C 0　　　　　　　　　　**D** 5

8 Which number is greater than six and less than nine?

　A five　　　　　　　　　**B** seven

　C ten　　　　　　　　　 **D** three

9 Which one has an answer greater than 8?

A 5 + 2

B 1 + 7

C 5 + 5

D 4 + 3

10 4 balls are in the bucket.
5 balls are in the basket.
How many balls are there in all?

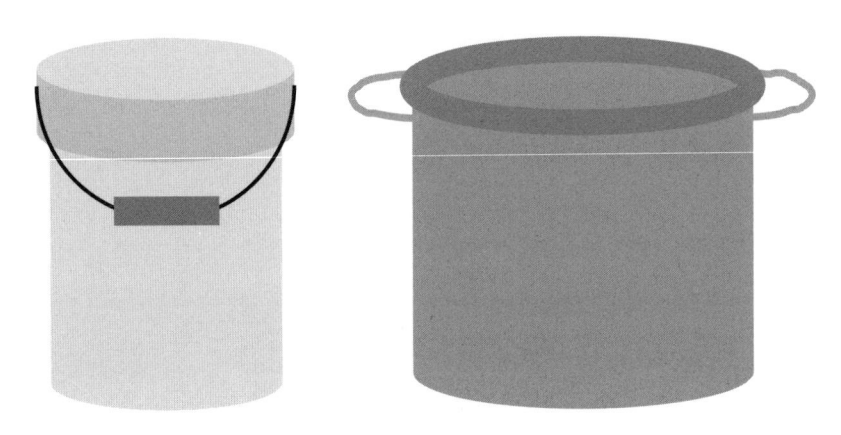

A 9

B 2

C 5

D 8

Section B (2 points each)

11 Write the missing numbers.

| 10 | 9 | | | 6 | |

12 Check (✓) the set that has one more than eight.

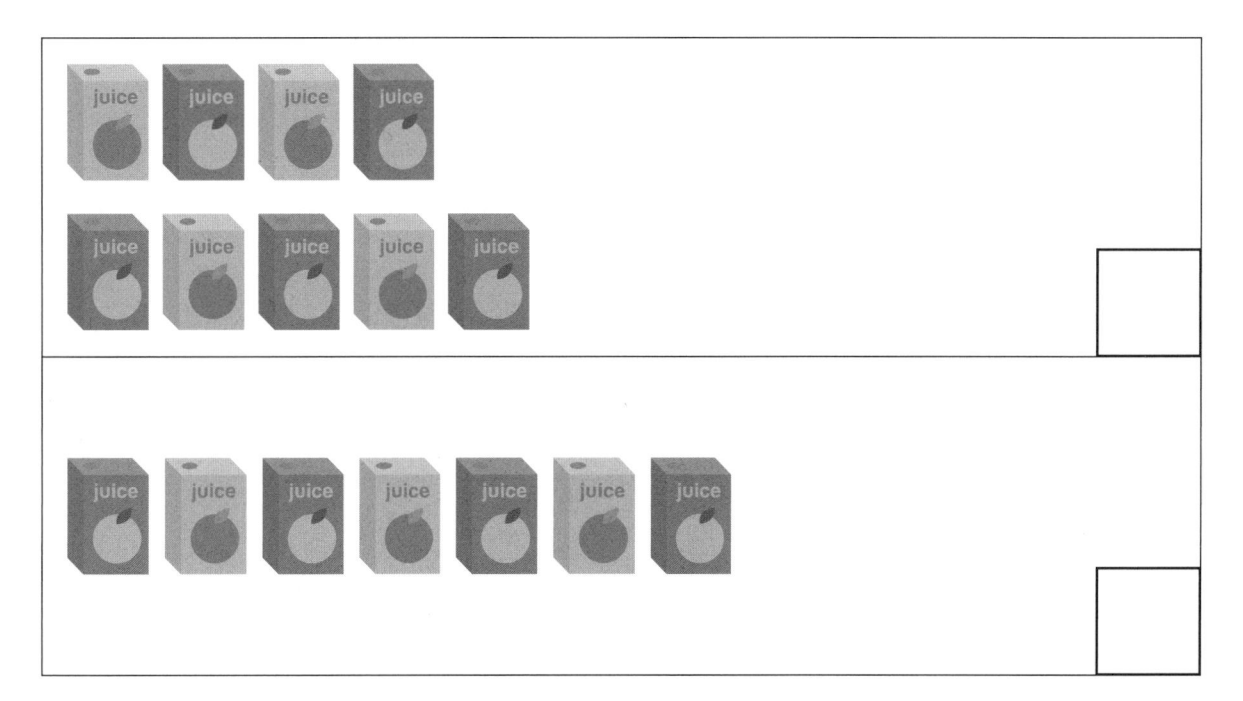

13 Write the numbers in order from least to greatest.

5   0   6
  8   10

Continual Assessment 1 Test B

14 Cross out the number that is 2 more than 4 − 3.

| 4 | 3 | 2 | 5 |

15 _____ is more than 8 and less than 10.

16 Complete the number bonds.

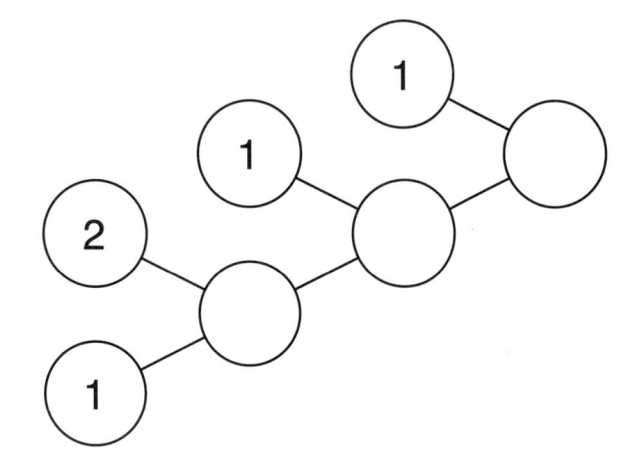

17 Use these numbers to make 2 different number bonds.

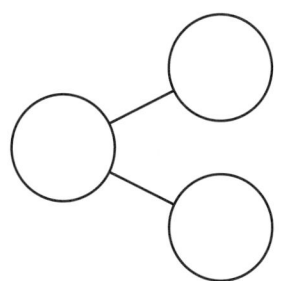

| 3 | 2 | 5 | 7 |

18 Circle the 2 numbers that together make 9.

| 2   8   4   6   5 |

19 Write 2 different addition equations.

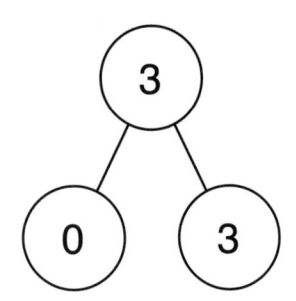

```
    3
   / \
  0   3
```

0 + ☐ = ☐

☐ + ☐ = ☐

20 Write 2 different subtraction equations.

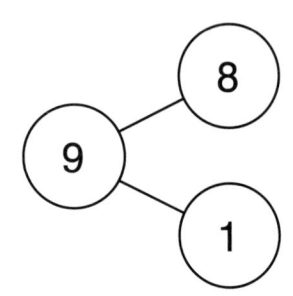

```
     8
    /
   9
    \
     1
```

☐ − 8 = ☐

☐ − ☐ = ☐

Section C (4 points each)

21 There are 3 owls on a branch.
There are 2 owls on another branch.
How many owls are there in all?

□ ○ □ = □

There are _____ owls in all.

22 There are 6 shells in a bucket.
David takes 1 shell out.
How many shells are still left in the bucket?

□ ○ □ = □

There are _____ shells still left in the bucket.

23 Kona has 3 small pumpkins and 5 big pumpkins.
How many pumpkins does she have altogether?

☐ ○ ☐ = ☐

Kona has _____ pumpkins altogether.

24 There are 4 ducks in the pond.
There are 2 ducks on the shore.
How many ducks are there in all?

☐ ○ ☐ = ☐

There are _____ ducks in all.

25 Sam has 10 pieces of fruit.
7 are apples.
The rest are oranges.
How many oranges does Sam have?

☐ ○ ☐ = ☐

Sam has _____ oranges.

Name: _____

Date: _____

25 min   Score

30

Test A

Chapter 5 Numbers to 20

Section A (2 points each)

Circle the correct option: **A**, **B**, **C**, or **D**.

1 What does 10 and 2 make?

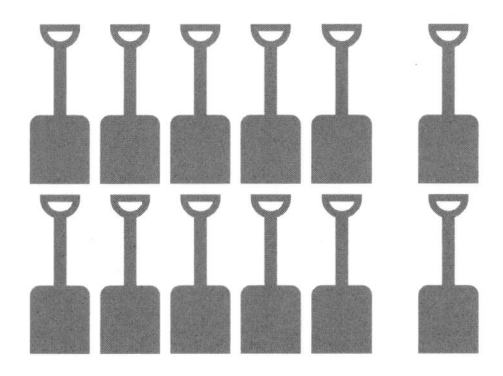

A 5 **B** 12

C 10 **D** 2

2 How many fish are there altogether?

A 14 **B** 10

C 15 **D** 16

3 Which one is the greatest number?

A 19

B 12

C 10

D 14

4 17 + 2 is _____.

A 19

B 14

C 17

D 12

5 16 − 3 is _____.

A eight

B fifteen

C eighteen

D thirteen

Section B (2 points each)

6 Write the missing numbers.

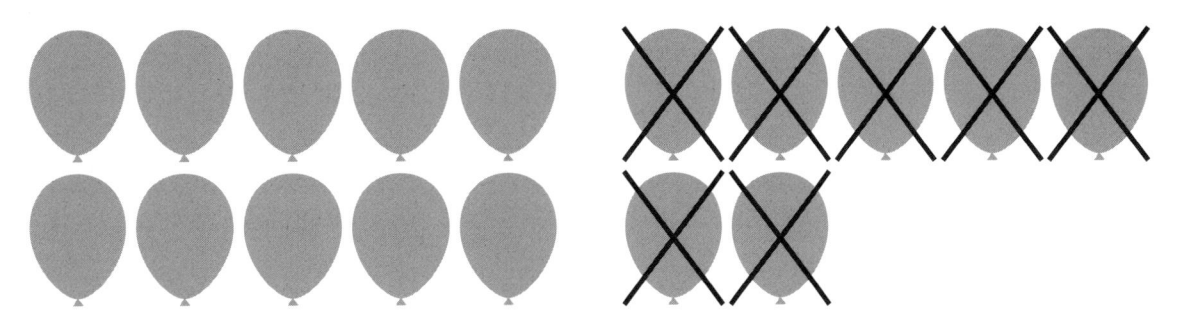

☐ – 7 = ☐

7 Check (✓) the set that has one more than 15.

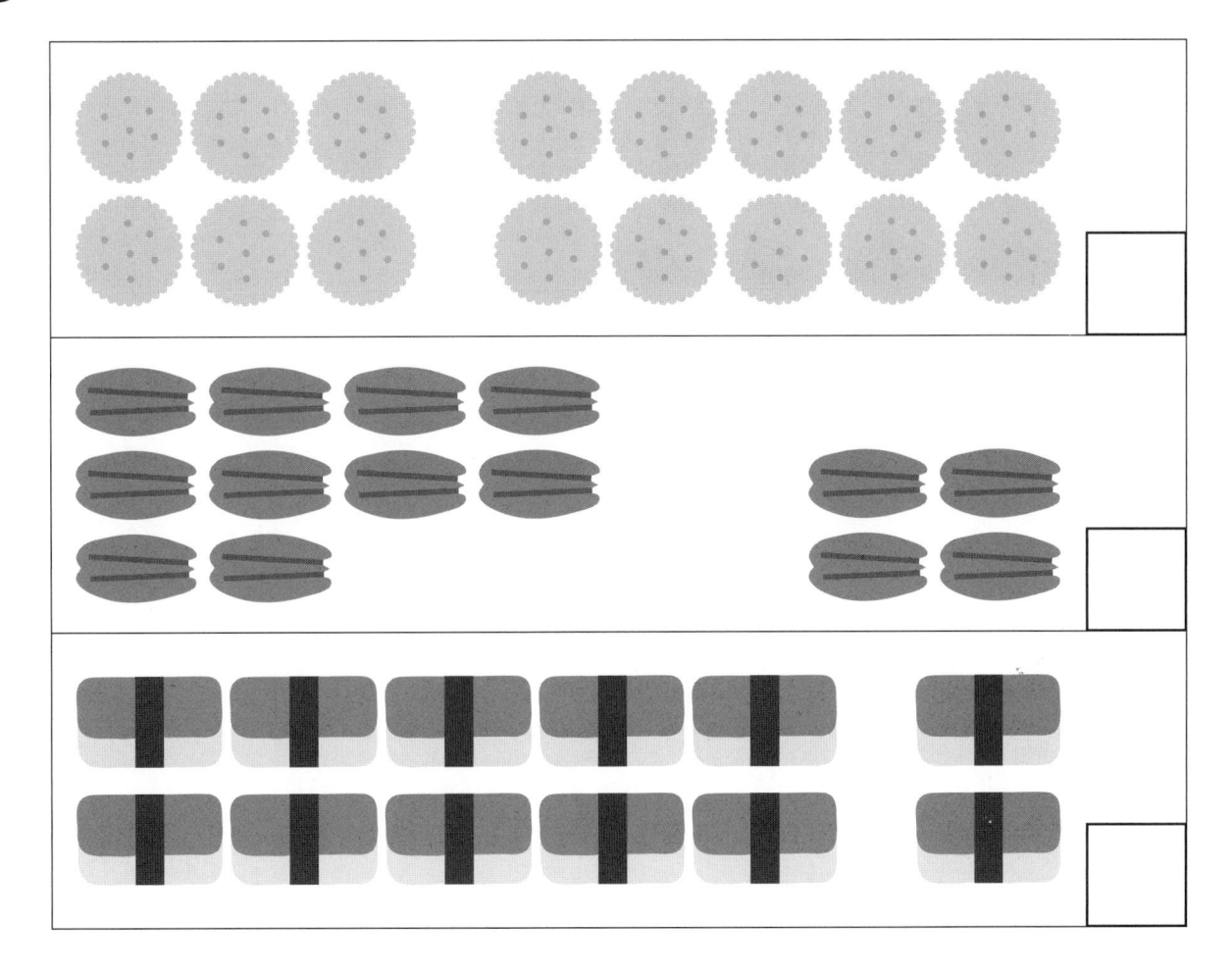

8 Cross out the number that is 2 more than eleven.

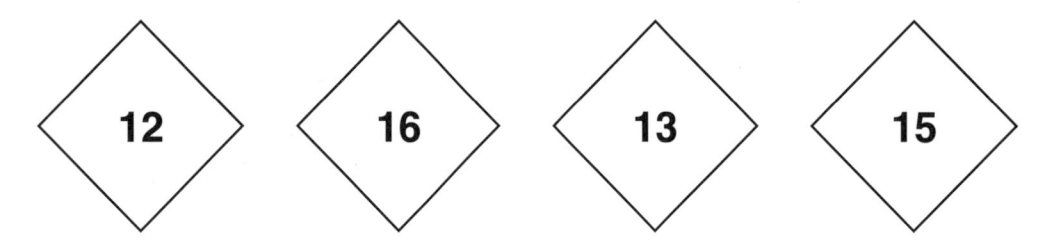

⟨ 12 ⟩   ⟨ 16 ⟩   ⟨ 13 ⟩   ⟨ 15 ⟩

9 Check (✓) the box that has fewer.

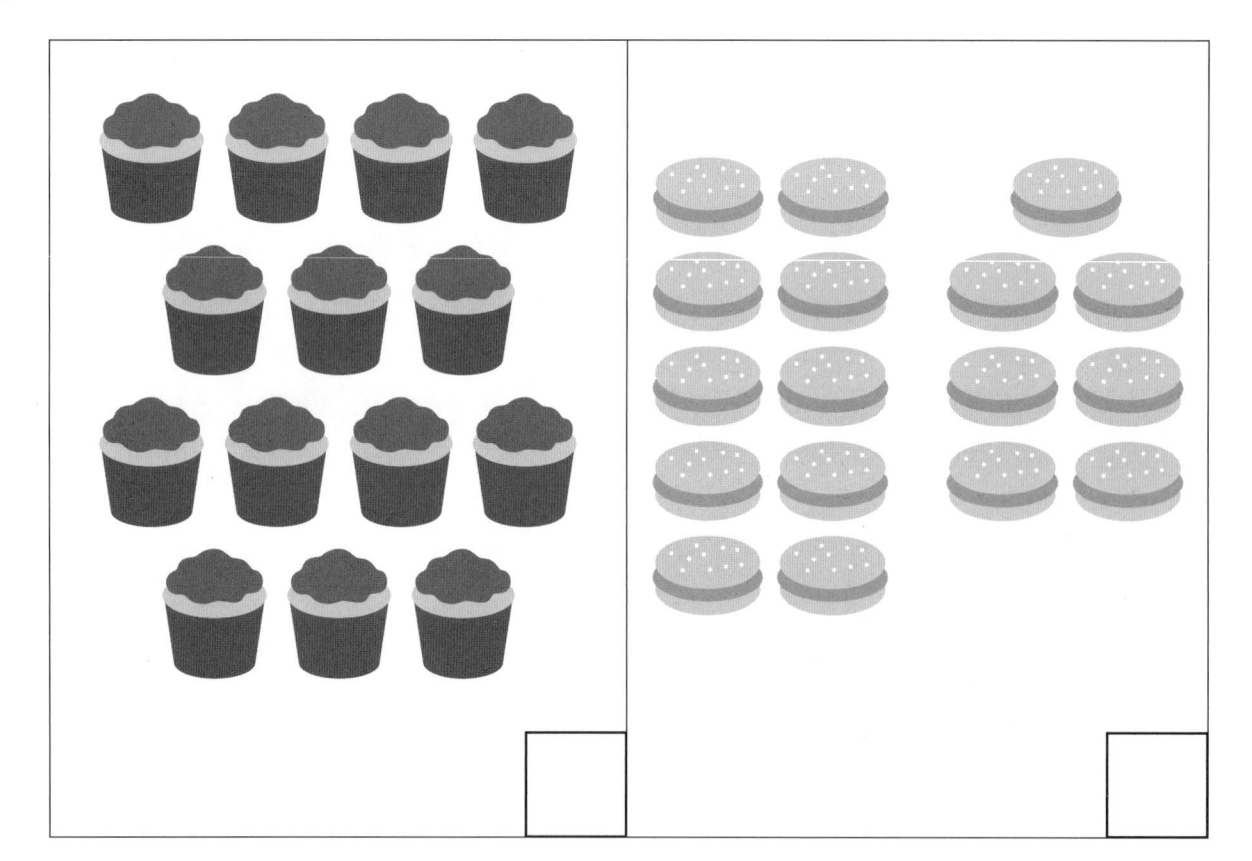

10 Write the numbers in order, from least to greatest.

| 11 17 19 14 |

_____ , _____ , _____ , _____
least   greatest

11 Circle the two numbers that are greater than 15.

13   20

14   12   16

12 Complete the number pattern.

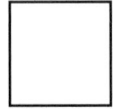

| 18 | | 16 | | | 13 |

Chapter 5 Test A

73

13 Write the number that is two more than eighteen.

☐

14 Kona has 10 beads.
Ella has 3 more beads than Kona.
How many beads does Ella have?

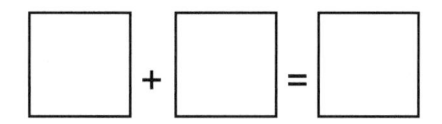

☐ + ☐ = ☐

Ella has _____ beads.

15 There are 15 birds in a tree.
4 fly away.
How many birds are still in the tree?

☐ ◯ ☐ = ☐

_____ birds are still in the tree.

Name: _____

Date: _____

25 min   Score

30

Test B

Chapter 5 Numbers to 20

Section A (2 points each)
Circle the correct option: **A**, **B**, **C**, or **D**.

1 15 and _____ make 17.

A 5

B 7

C 3

D 2

2 Which number is less than fourteen?

A twenty

B twelve

C sixteen

D eighteen

3 Which one gives the same answer as 10 + 6?

    A 11 + 4   **B** 16 + 1

    C 18 − 2   **D** 13 − 3

4 In which set are the numbers arranged in order from least to greatest?

    A | 11, 13, 20, 15 | **B** | 12, 14, 18, 20 |

    C | 14, 16, 12, 20 | **D** | 19, 17, 15, 13 |

5 Which equation is false?

    A 13 = 17 − 4   **B** 20 = 19 + 1

    C 12 = 15 − 2   **D** 16 = 11 + 5

Section B (2 points each)

6 Check (✓) the box that has more than 15.

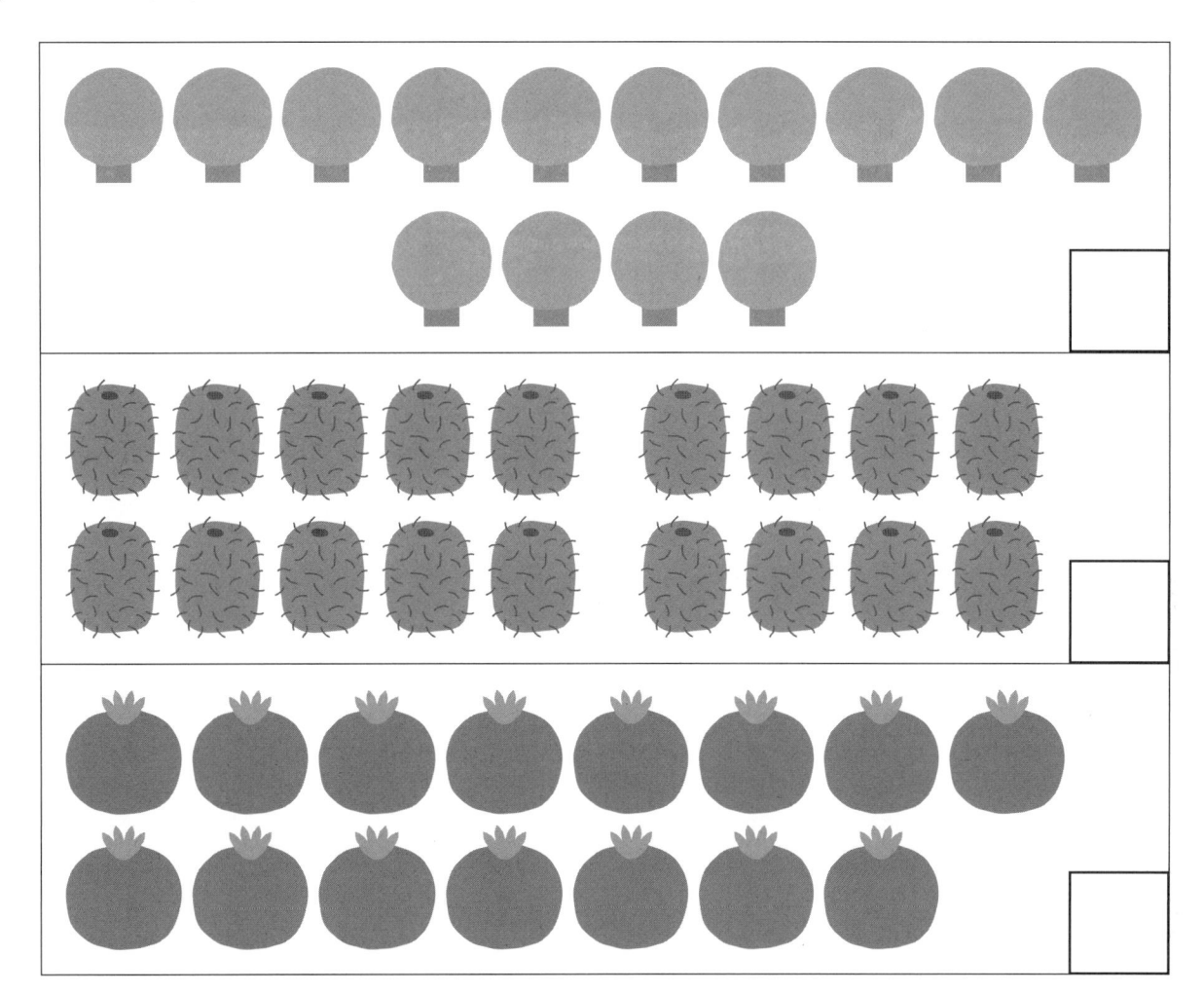

7 Write the missing number.

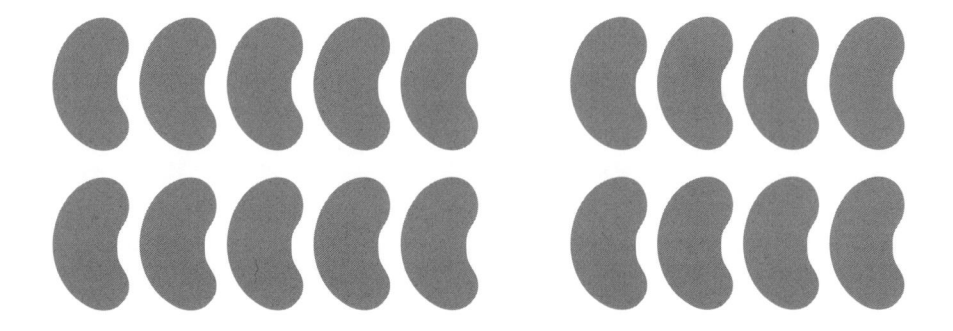

5 less than 18 is ☐ .

Chapter 5 Test B

8 Complete the number bond.

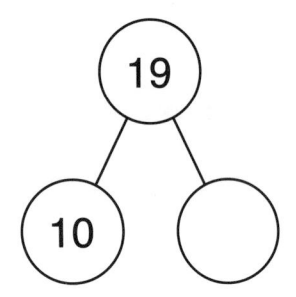

```
    19
   /  \
  10   ◯
```

9 Write the missing number.

☐ − 3 = 14

10 Write the missing number.

11 + 4 = ☐ − 4

11 Complete the number pattern.

| 6 | 8 | | 12 | | |

12 Write the numbers in order, from least to greatest.

| 10 | 20 | 12 | 18 | 17 |

| | | | | |

13 Circle the two numbers that are less than 17 and greater than 14.

11   15

16   20   19

Chapter 5 Test B   79

14 Daren has 15 crayons.
Chapa has 5 fewer crayons than Daren.
How many crayons does Chapa have?

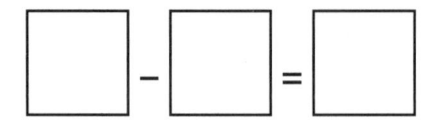

☐ − ☐ = ☐

Chapa has _____ crayons.

15 Cooper has 4 more red apples than green apples.
He has 13 green apples.
How many red apples does he have?

☐ ◯ ☐ = ☐

He has _____ red apples.

Name: _____

Date: _____

25 min   Score

30

Test A

Chapter 6 Addition to 20

Section A (2 points each)
Circle the correct option: **A**, **B**, **C**, or **D**.

1 There are 6 big apples and 7 small apples. How many apples are there in all?

A 15   **B** 13

C 16   **D** 17

2 8 and _____ is 13.

A 13   **B** 8

C 5   **D** 3

3 5 more than 9 is _____.

    A 12   **B** 13

    C 14   **D** 15

4 Which of the following gives an answer of 12?

    A 6 + 6   **B** 11 + 2

    C 6 + 5   **D** 4 + 9

5 Which number is greater than 14 + 3?

    A 14   **B** 16

    C 18   **D** 13

Section B (2 points each)

6 Circle the boxes that give an answer of 11.

| 9 and 3 | 11 and 0 | 5 and 8 | 4 and 7 |

7 Write the missing number.

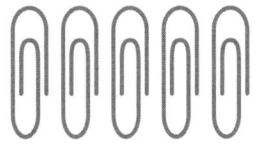

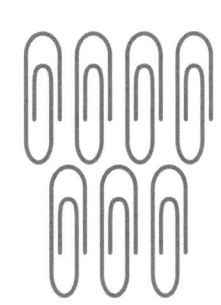

5 + 7 = ☐

8 Circle the number that is 7 more than 6.

| 7   13   16   10 |

Chapter 6 Test A

9 8 + 8 = ☐

10 4 + 9 = 10 + ☐
  ◯ ①

11 Circle the cards that have the answer of 15.

| 11 + 5 | 6 + 9 | 8 + 5 | 7 + 8 |

12 Complete the equation.

8 + ☐ = 12

13 There are 6 sand dollars and 5 starfish. How many sea animals are there in all?

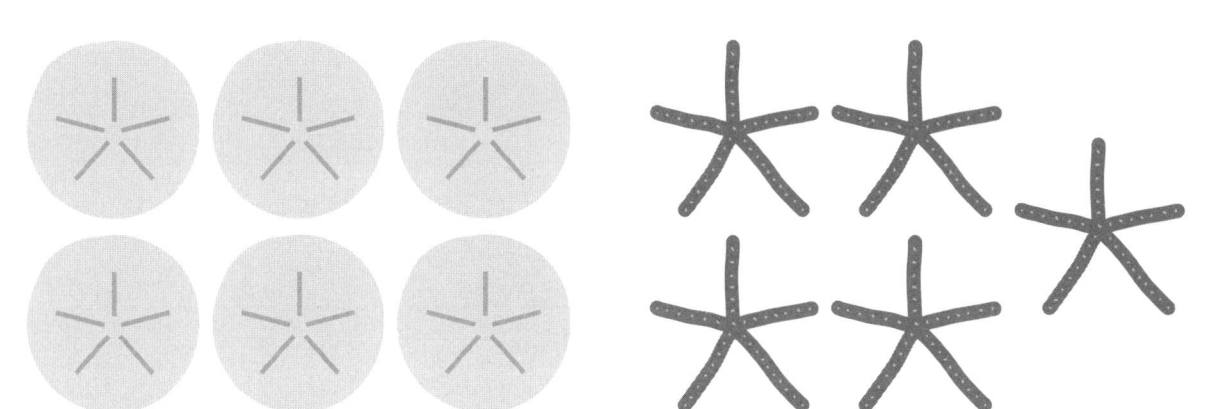

☐ + ☐ = ☐

14 How many candies are there in all?

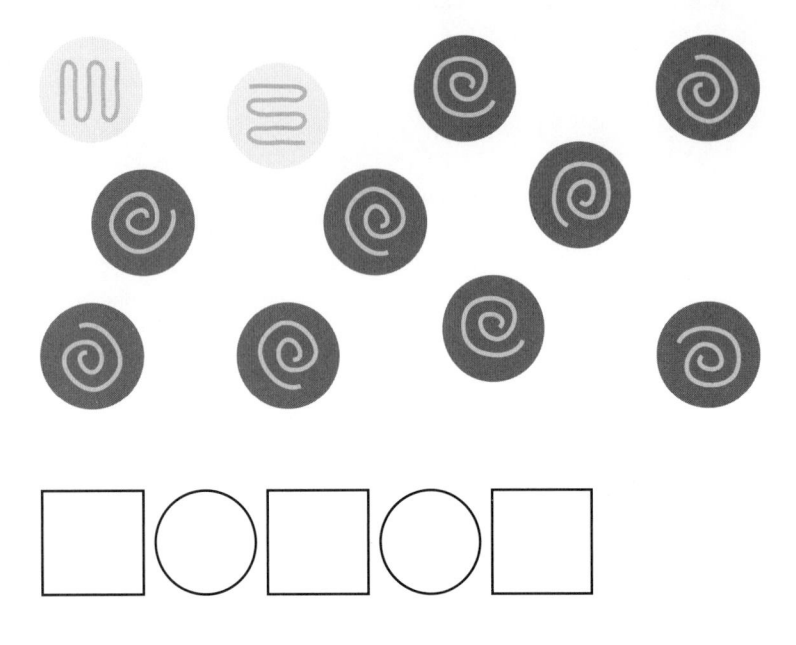

There are _____ candies in all.

15 There are 6 dogs and 6 cats.
How many animals are there altogether?

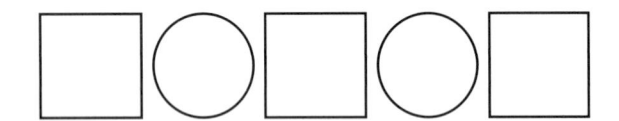

There are _____ animals altogether.

Name: _____

Date: _____

25 min  Score

30

Test B

Chapter 6 Addition to 20

Section A (2 points each)
Circle the correct option: **A**, **B**, **C**, or **D**.

1 Which picture shows 15 cookies?

A

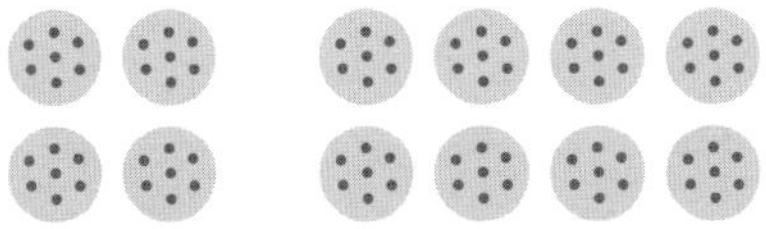

B

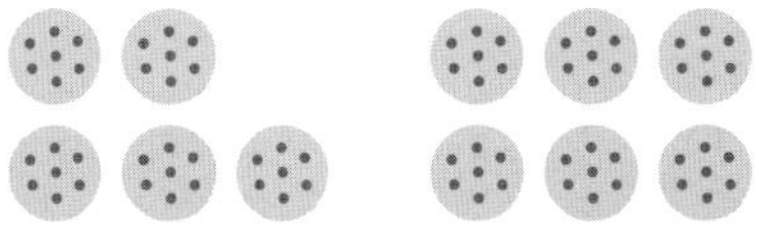

C

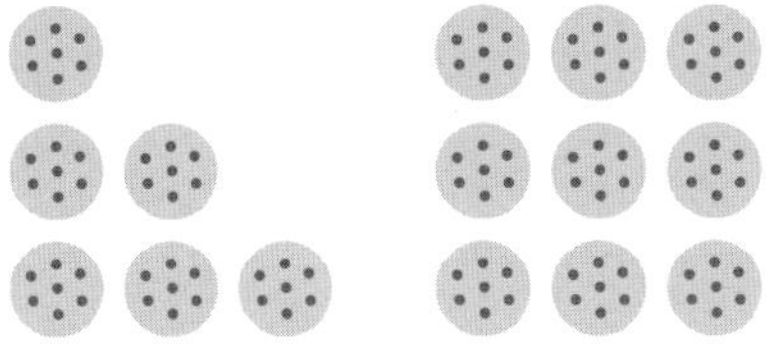

D

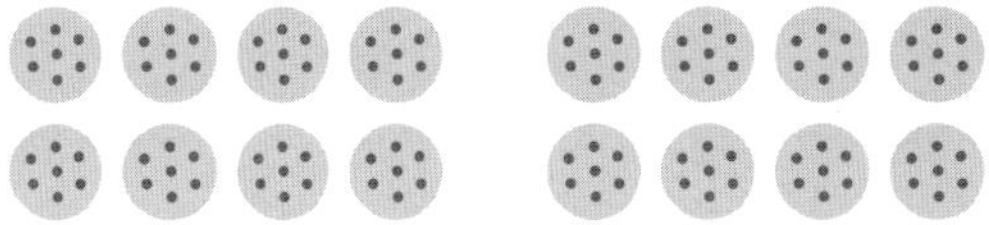

2 Add 5 to 12.
What is the answer?

A 15

B 7

C 18

D 17

3 Which number is 7 more than 4?

A 11

B 10

C 14

D 17

4 Which number is less than 6 + 8?

A 17

B 16

C 18

D 13

5 Which number is 1 greater than 14 + 3?

A twelve

B nineteen

C eighteen

D six

Section B (2 points each)

6 Cross out the box with the least answer.

| 7 + 5 | | 9 + 6 | | 2 + 9 |

7 Circle the box that gives a number that is greater than fourteen.

| 7 + 7 | | 6 + 9 | | 8 + 4 |

8 Write the missing number.

4 + 6 = 1 + ☐

9 Write the missing number.

8 + 4 = 7 + ☐

10 4 more than 13 is the same as 5 more than _____.

11 Check (✓) the equation that is true.

| 13 + 3 = 13 | ☐ |

| 6 + 9 = 15 | ☐ |

| 18 = 8 + 8 | ☐ |

12 Circle the box that gives an answer greater than 4 + 9.

| 7 + 4 | 8 + 6 | 0 + 13 | 5 + 6 |

13 There are 4 fish in a fish tank.
Mei adds 12 more fish.
How many fish are there altogether?

☐ + ☐ ◯ ☐

There are _____ fish altogether.

14 Koni buys 7 oranges and 8 apples.
How many oranges and apples does she buy in all?

☐ ○ ☐ ○ ☐

She buys _____ oranges and apples in all.

15 Alex has 4 stickers.
Mei gives him 9 more.
How many stickers does Alex have altogether?

☐ ○ ☐ ○ ☐

Alex has _____ stickers altogether.

Name: _____

Date: _____

25 min

Score

30

Test A

Chapter 7 Subtraction Within 20

Section A (2 points each)
Circle the correct option: **A**, **B**, **C**, or **D**.

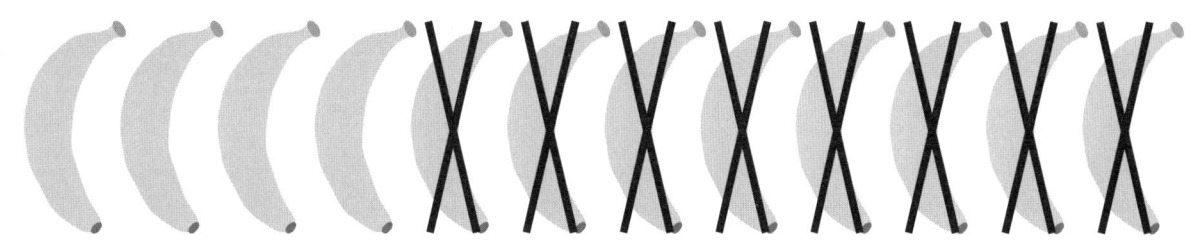

1

12 − 8 = ☐

A 4

B 8

C 12

D 2

2 4 less than 15 is _____.

A 10

B 11

C 16

D 5

3 There are 14 bunnies in the garden.
6 bunnies run away.
How many bunnies are left?

A 6

B 10

C 4

D 8

4 What number is 9 less than 18?

A 9

B 16

C 8

D 10

5 What is the missing number?

13 − ☐ = 7

A 6

B 5

C 11

D 9

94 Chapter 7 Test A

Section B (2 points each)

6 Subtract 5 from 13.

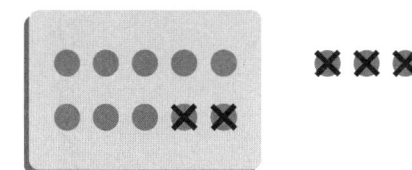

13 − 5 = ☐

7 Write the missing number.

16 − 9 = ☐

8 Write the missing number.

13 − ☐ = 4

9 Circle the number that is equal to 11 − 5.

| 9 | 14 | 16 | 6 |

10 6 less than 12 is _____.

11 Circle the box that has an answer of 8.

| 11 − 4 | | 16 − 5 | | 18 − 8 | | 17 − 9 |

12 Complete the equation.

☐ − 4 = ☐

13 How many slices of watermelon are left?

☐ − ☐ = ☐

There are _____ slices of watermelon left.

Chapter 7 Test A

14 There are 15 crabs.
7 crabs crawl away.
How many crabs are left?

☐ ○ ☐ = ☐

_____ crabs are left.

15 12 children are on a bus.
Some children get off the bus.
9 children are still on the bus.
How many children get off the bus?

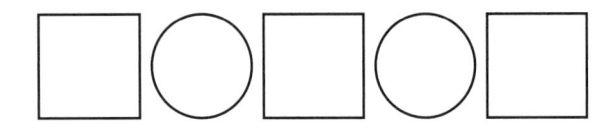

☐ ○ ☐ ○ ☐

_____ children get off the bus.

Name: _____

Date: _____

25 min   Score

30

Test B

Chapter 7 Subtraction Within 20

Section A (2 points each)
Circle the correct option: **A**, **B**, **C**, or **D**.

1 4 less than 11 is _____.

  A 4   **B** 7

  C 11   **D** 15

2 A baker has 15 cupcakes.
He sells 6 of them.
He has _____ cupcakes left.

  A 9   **B** 15

  C 13   **D** 5

3 Which one gives an answer of 5?

A 11 – 9

B 13 – 5

C 17 – 2

D 11 – 6

4 Laila has 16 crayons.
She gives some crayons away.
She has 9 crayons left.
She gives away _____ crayons.

A 7

B 10

C 4

D 8

5 What number is 1 less than 18 – 9?

A 8

B 9

C 17

D 10

Section B (2 points each)

6 What is 2 less than 17?
Circle the correct answer.

| twelve | fourteen | fifteen | seventeen |

7 Circle the boxes that have an answer of 8.

| 14 – 6 | 17 – 8 | 15 – 7 | 14 – 9 |

8 Cross out the box with an equation that is false.

| 11 – 9 = 8 | 16 – 9 = 7 |

9 Write the missing number.

☐ − 6 = 7

10 Complete the equation.

14 − 7 = 12 − ☐

11 Complete the equation.

15 − ☐ = 6 + 3

12 Ella buys 11 apples.
She eats 2 of them.
How many apples does she have left?

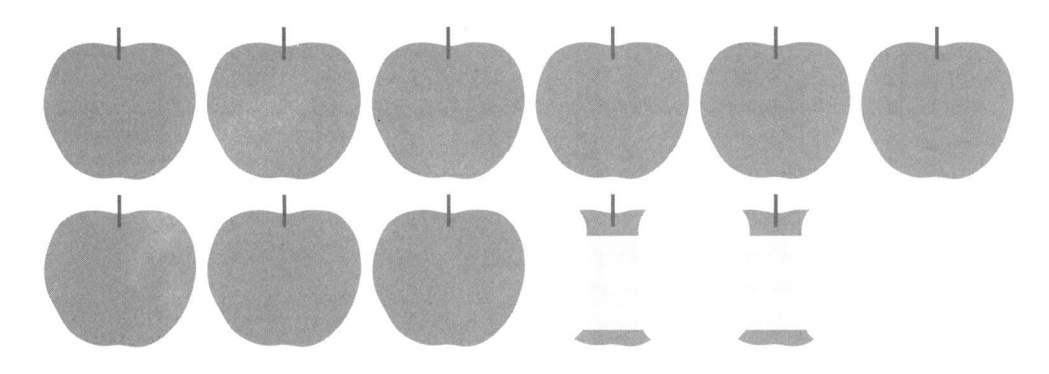

☐ − ☐ = ☐

She has _____ apples left.

13 Mei has 5 fish in her fish tank.
How many does she need to add to have 12 fish in the fish tank?

12 − ☐ = ☐

She needs to add _____ fish.

Chapter 7 Test B 103

14 Dion has 14 stickers.
He uses 3 of them.
How many stickers does he have left?

☐○☐○☐

Dion has _____ stickers left.

15 Jacob has 15 books on a bookshelf.
How many books does he need to remove to have 7 books on the bookshelf?

☐○☐○☐

Jacob needs to remove _____ books.

Name: _____

Date: _____

25 min   Score

30

Test A

Chapter 8 Shapes

Section A (2 points each)
Circle the correct option: **A**, **B**, **C**, or **D**.

1 How many sides does this shape have?

A 5   **B** 3

C 6   **D** 4

2 Which shape has 4 corners?

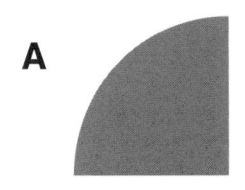

A

B

C

D

Chapter 8 Test A   105

3 Which shape has only 3 sides?

A (star)

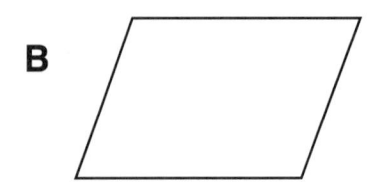

B (parallelogram)

C (right triangle-like shape)

D (triangle)

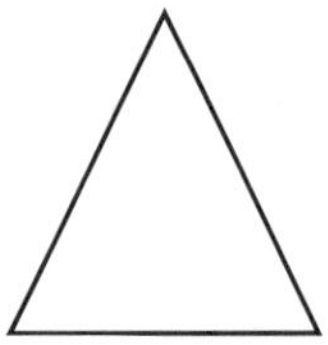

4 What comes next in the pattern?

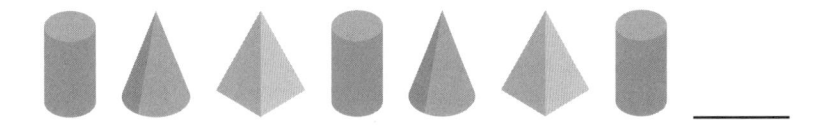

(cylinder, cone, pyramid, cylinder, cone, pyramid, cylinder, ___)

A (pyramid)

B (cylinder)

C (cone)

D (cube)

5 What shape is missing in the pattern?

(triangle, upside-down triangle, circle, small circle, triangle, upside-down triangle, circle, small circle, triangle, ?, circle, small circle)

A (small circle)

B (triangle)

C (circle)

D (upside-down triangle)

106 Chapter 8 Test A

Section B (2 points each)

6 Circle the squares.

7 Cross out the shape that does not belong.

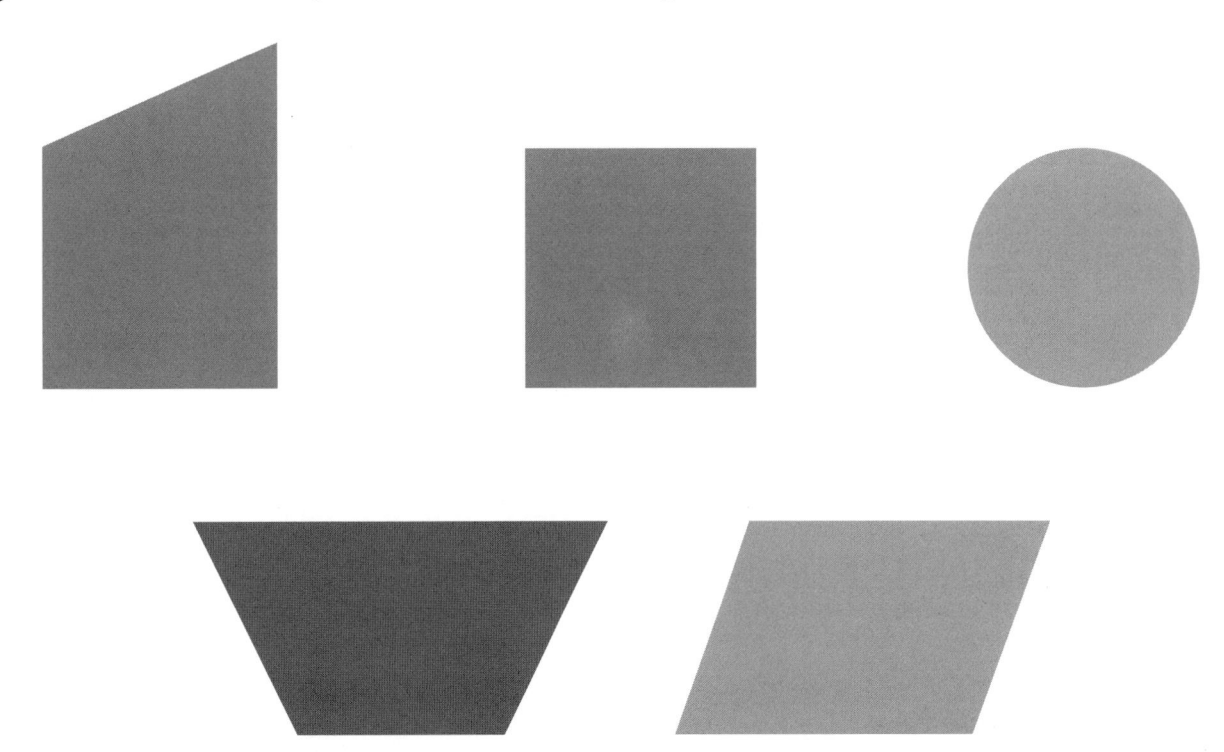

Chapter 8 Test A

8 Check (✓) the shape that has more than 4 corners.

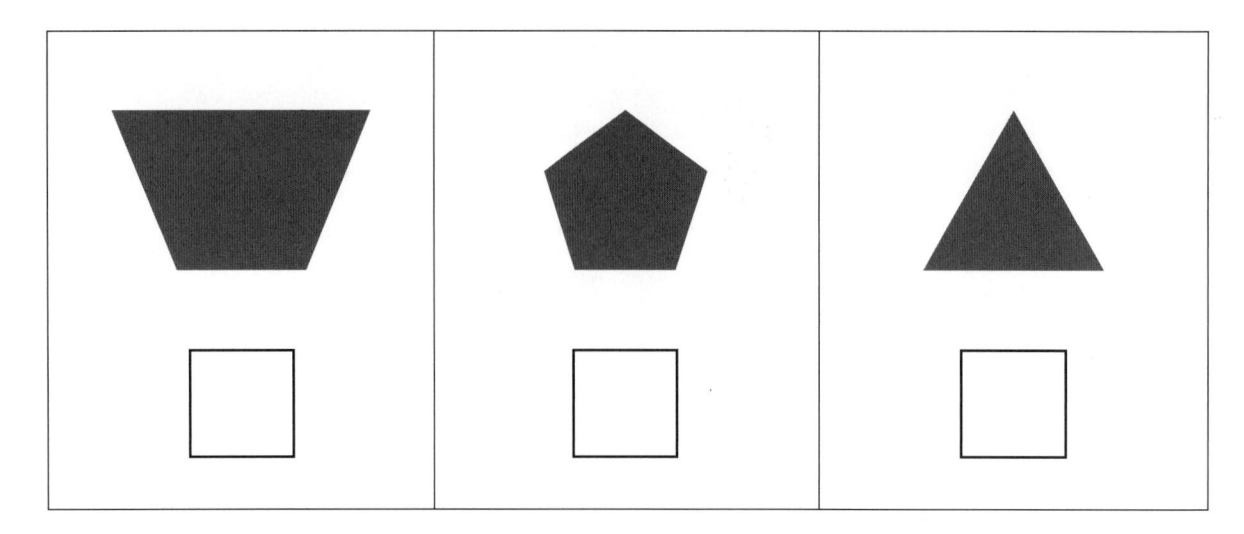

9 How many circles does this picture have?

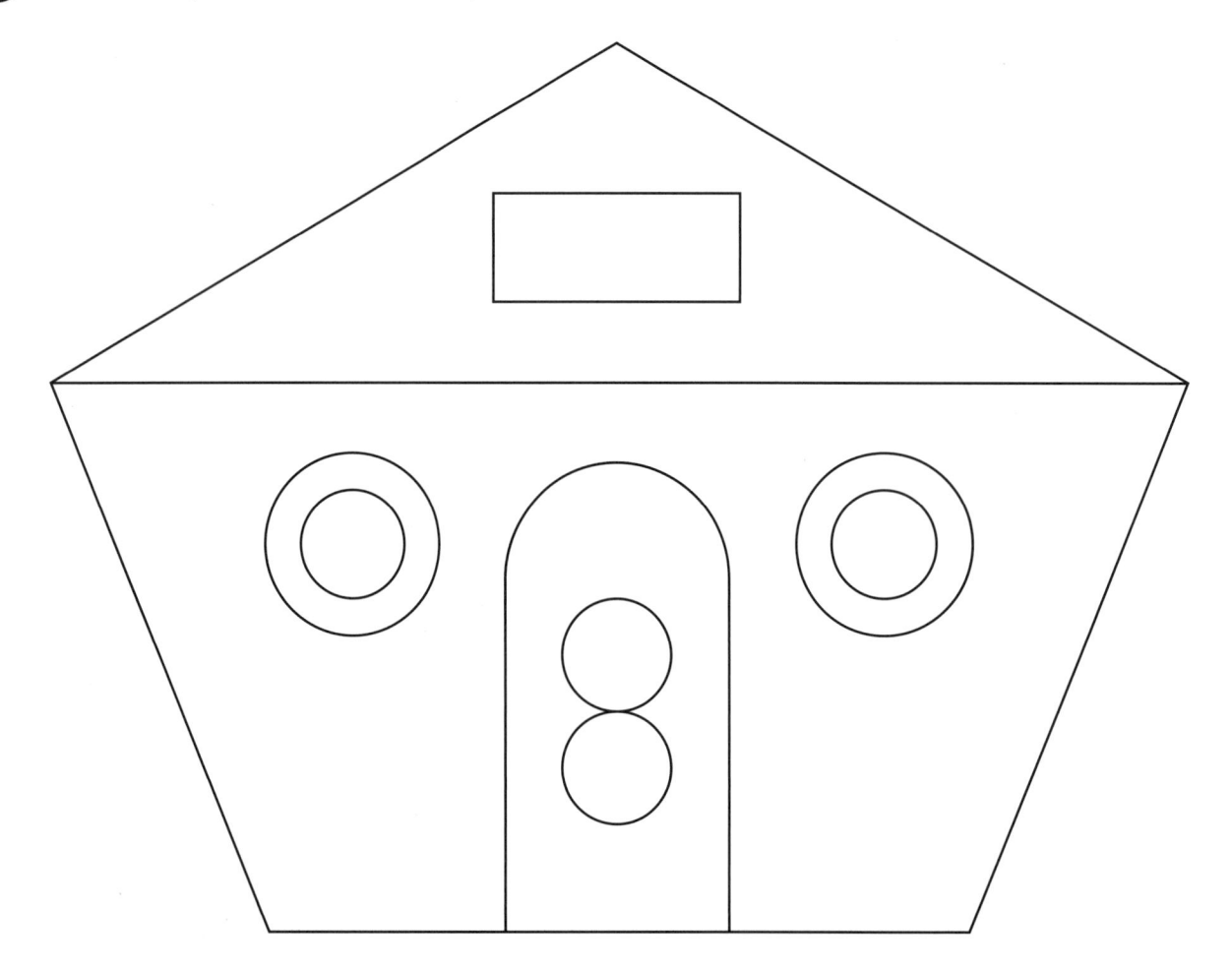

This picture has _____ circles.

108 Chapter 8 Test A

10 Cross out the shapes that do not have 4 sides.

11 Color all the shapes that have only 3 corners.

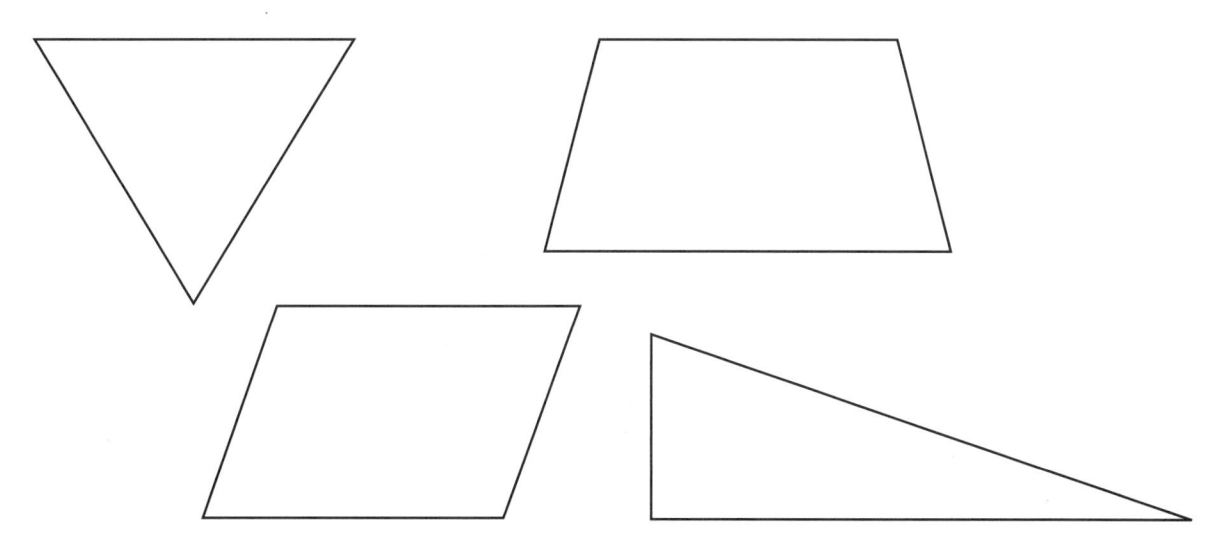

12 Circle the shape that comes next in the pattern.

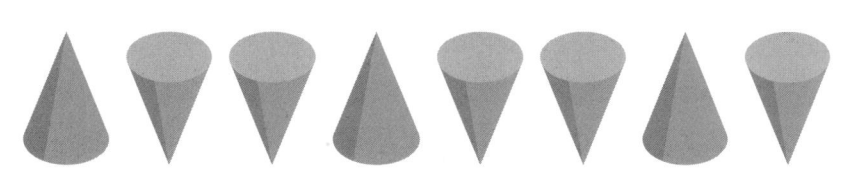

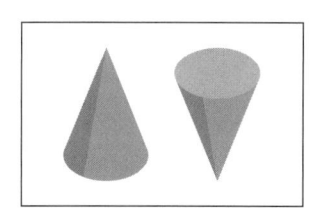

Chapter 8 Test A

13 What comes next?
Draw it.

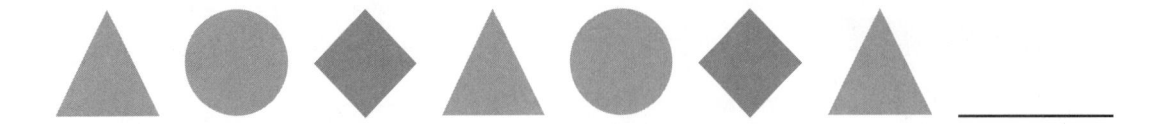

△ ● ◆ △ ● ◆ △ ____

14 Draw 1 line to cut this figure into two triangles.

15 Circle 2 figures that when put together will form a circle.

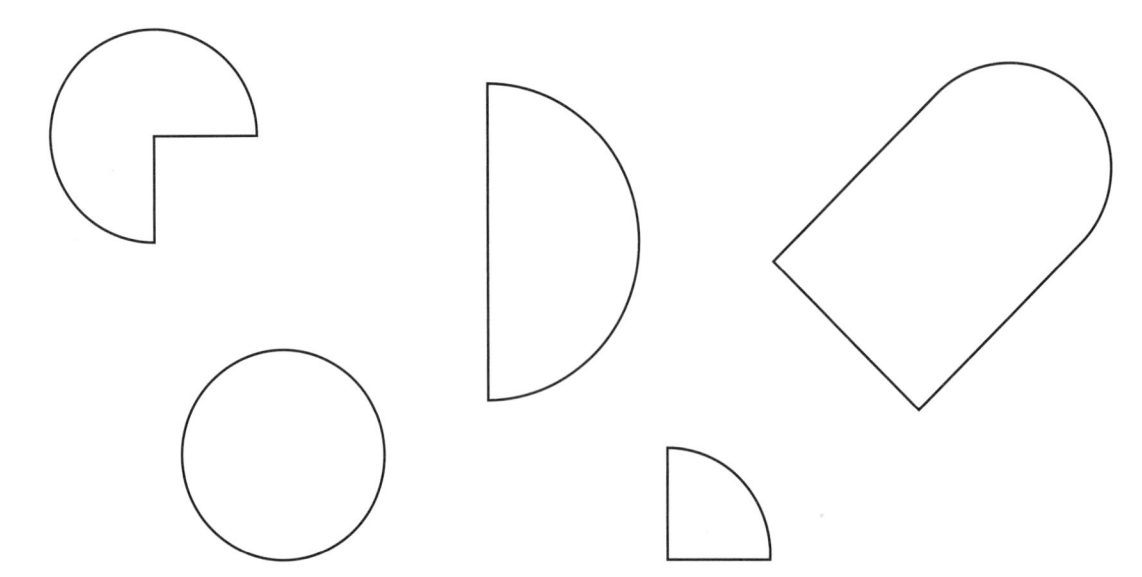

Name: _____

Date: _____

25 min   Score

30

Test B

Chapter 8 Shapes

Section A (2 points each)
Circle the correct option: **A**, **B**, **C**, or **D**.

1 How many sides does this shape have?

A 4   **B** 5

C 6   **D** 0

2 Which shape has more than 5 corners?

A

B

C

D

Chapter 8 Test B

111

3 A _____ has 3 sides and 3 corners.

A triangle

B circle

C rectangle

D square

4 Which shape is not in this pattern?

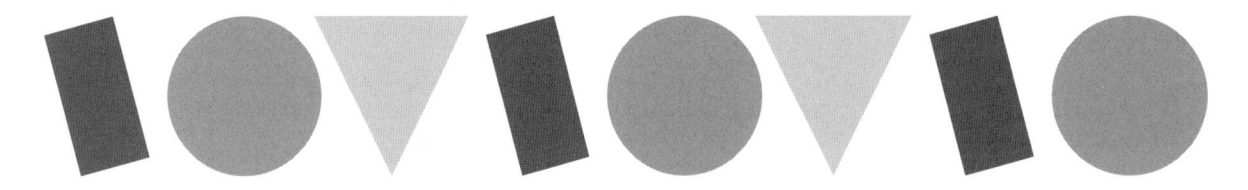

A rectangle

B circle

C triangle

D square

5 What shape comes next in this pattern?

A square

B rectangle

C circle

D triangle

Section B (2 points each)

6 Check (✓) the shape that has zero corners.

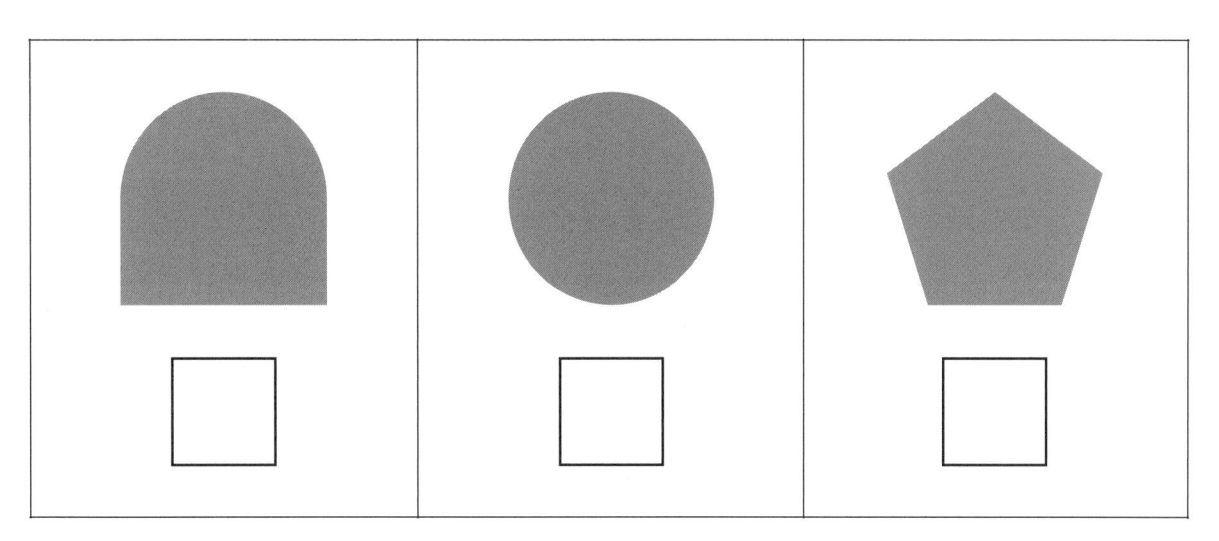

7 Draw a triangle in the square.

8 How many rectangles and triangles are in this picture?

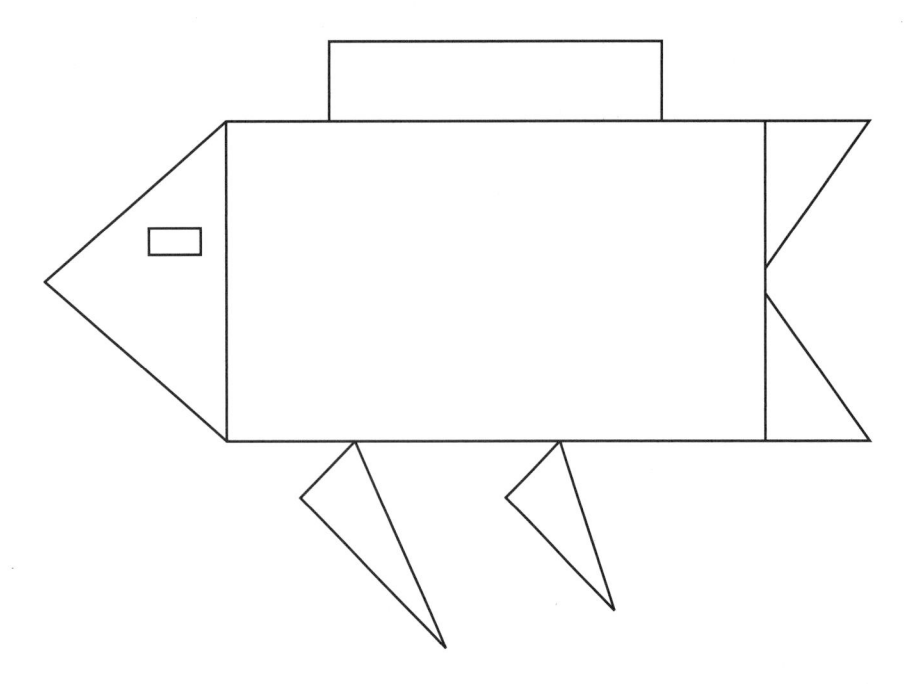

_____ rectangles   _____ triangles

9 A rectangle has _____ corners.

10 How many sides and corners does this figure have?

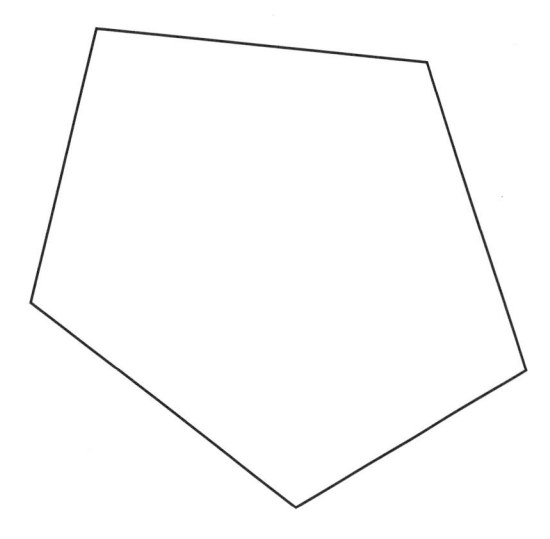

_____ corners   _____ sides

114   Chapter 8 Test B

11 Cross out the shape that does not belong.

12 What comes next in the pattern?
Circle it.

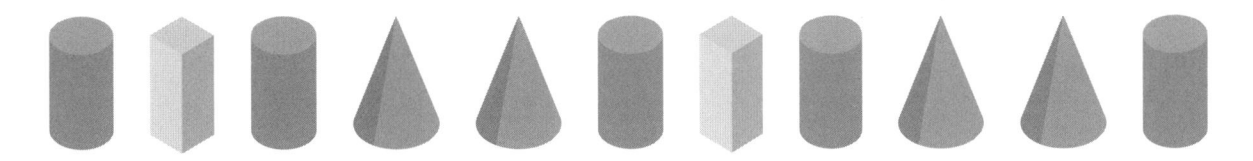

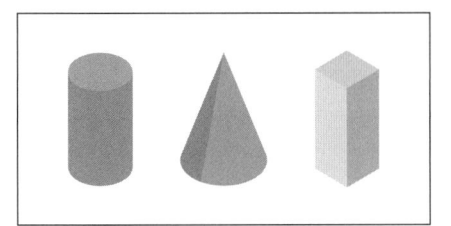

13 What comes next in the pattern?
Circle it.

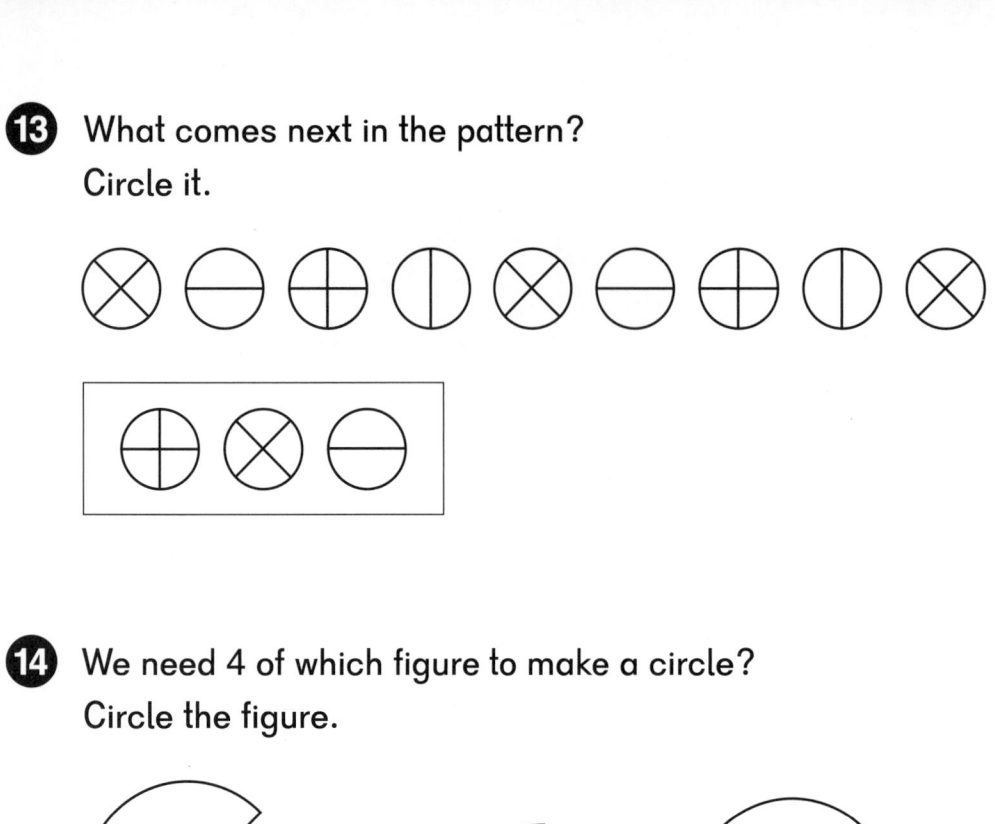

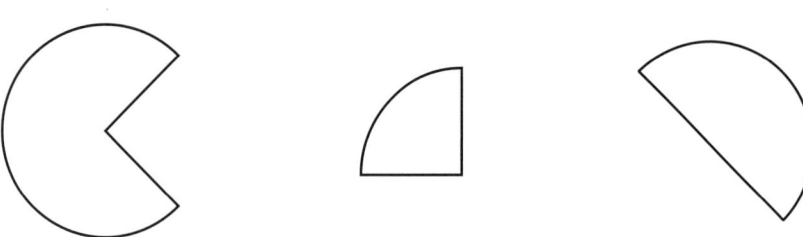

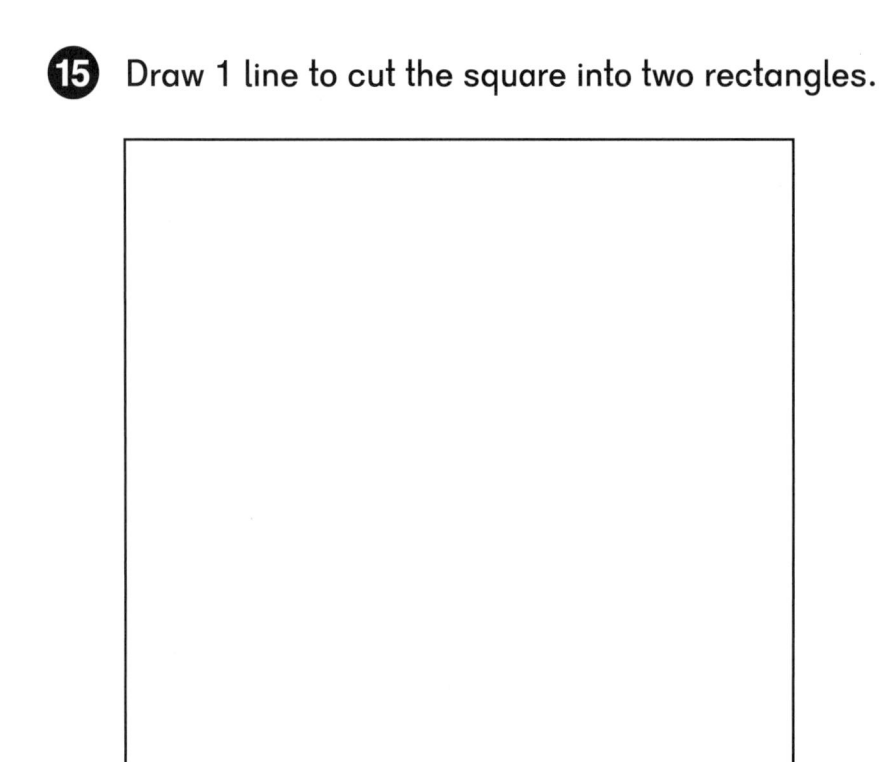

⊗ ⊖ ⊕ ⊘ ⊗ ⊖ ⊕ ⊘ ⊗

| ⊕ ⊗ ⊖ |

14 We need 4 of which figure to make a circle?
Circle the figure.

15 Draw 1 line to cut the square into two rectangles.

Name: _____

Date: _____

25 min   Score

30

Test A

Chapter 9 Ordinal Numbers

Section A (2 points each)
Circle the correct option: **A**, **B**, **C**, or **D**.

1 What position in line is the frog?

A 3rd

B 2nd

C 4th

D 5th

2 Which fruit is 3rd from the right?

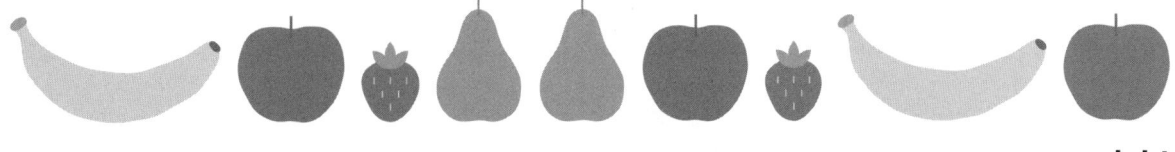

left   right

A banana

B apple

C strawberry

D pear

Chapter 9 Test A   117

3 How many ants are between the first and the last ant?

A 9

B 7

C 1

D 2

4 How many children are in line behind Mei?

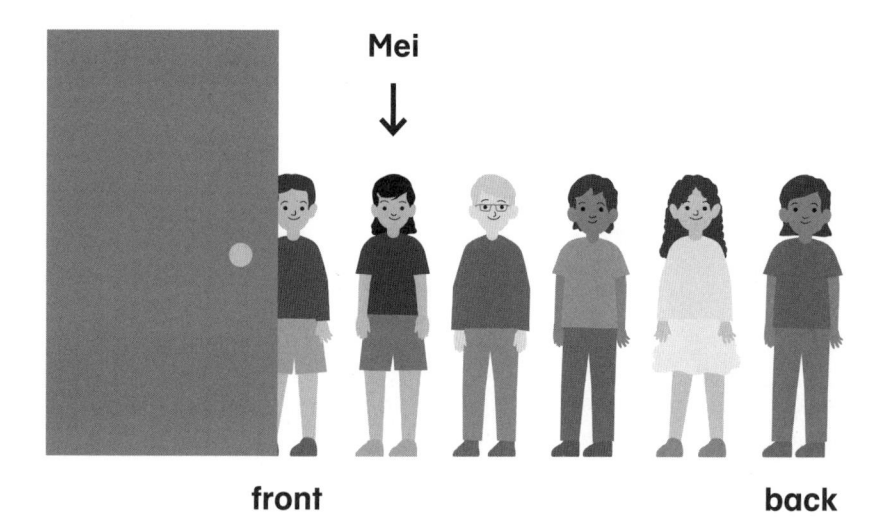

front

back

A 5

B 1

C 6

D 4

5 The butterfly is _____ from the left.

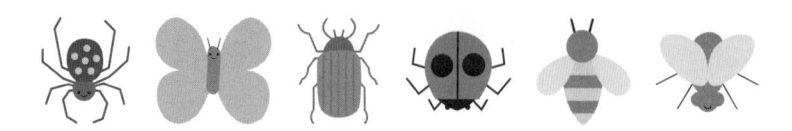

A 1st

B 6th

C 2nd

D 5th

118   Chapter 9 Test A

Section B (2 points each)

6 Laila's dog is the 5th dog from the right.
Which one is her dog?
Circle it.

left   right

7 Cross out the last 5 chicks.

1st

8 Draw an apple in the 4th square from the left.

left   right

9 Cross out the 2nd ball from the top.

10 Match the bird to the 1st nest.

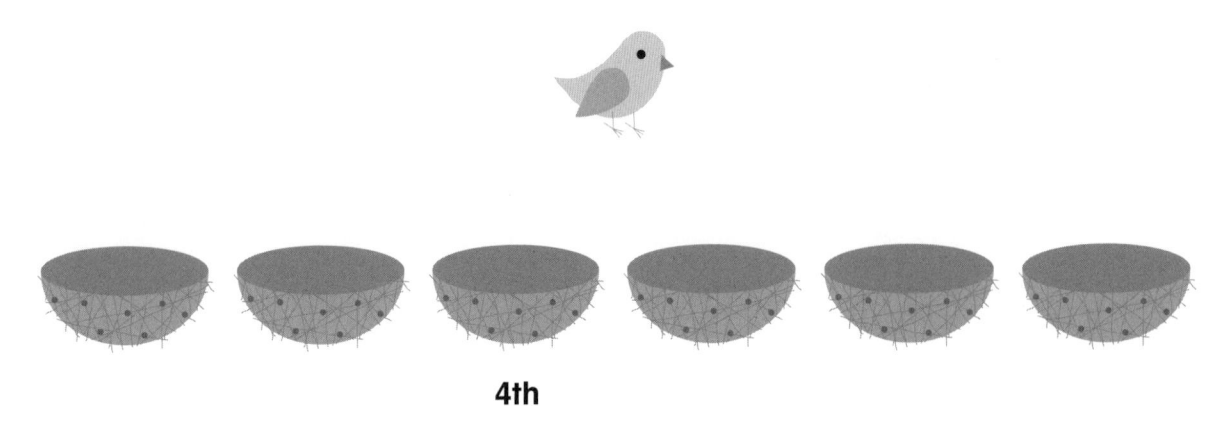

4th

11 How many cars are between the 2nd car and the 5th car?

There are _____ cars between the 2nd car and the 5th car.

12 Cross out the robot that is 2nd from the right and 7th from the left.

left　　　　　　　　　　　　　　　　　　　　　　　　　　**right**

13 Cross out the 6th box from the bottom.

Chapter 9 Test A

14 Dion is the 3rd person from the front of the line and the 3rd person from the back of the line. How many people are there in the line?

Dion

There are _____ people in the line.

15 Cody is in line waiting for the bus.
He is 6th in line.
There are 3 more people behind him.
How many people are in the line?

There are _____ people in the line.

Name: _____

Date: _____

25 min

Score: /30

Test B

Chapter 9 Ordinal Numbers

Section A (2 points each)
Circle the correct option: **A**, **B**, **C**, or **D**.

1 The monkey in the middle is _____ from the left.

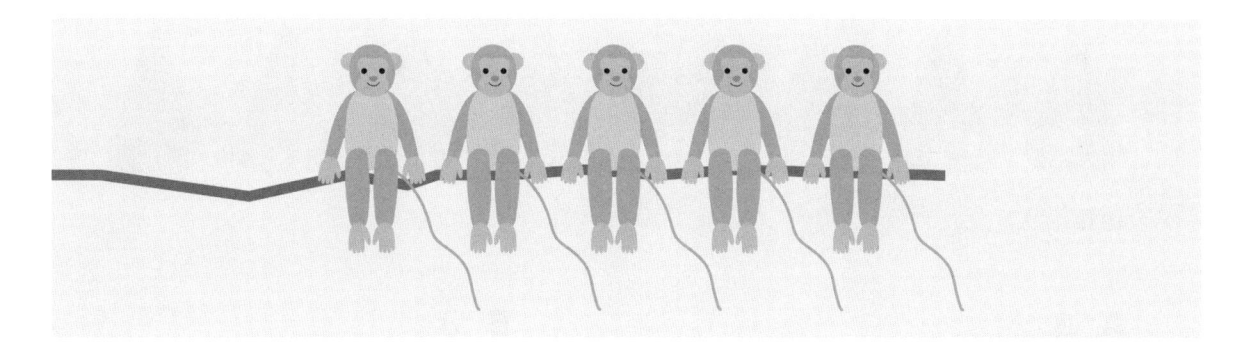

A 3rd

B 5th

C 1st

D 2nd

2 Which shape is in the 2nd and the 7th place from the right?

left   right

A triangle

B circle

C square

D rectangle

Chapter 9 Test B   123

3 How many rabbits are behind the 3rd rabbit?

first last

A 3 **B** 6

C 4 **D** 2

4 How many hats are to the left of the balloon?

left right

A 3 **B** 8

C 5 **D** 9

5 What is the 9th shape?

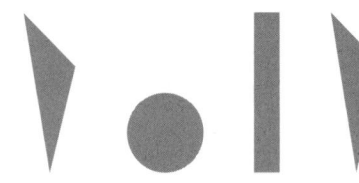

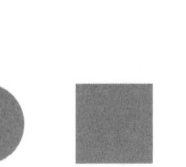

5th

A square **B** rectangle

C circle **D** triangle

Section B (2 points each)

6 Two animals are in front of the giraffe.
How many animals are behind the giraffe?

_____ animals are behind the giraffe.

7 Draw an apple on the 5th plate from the right.

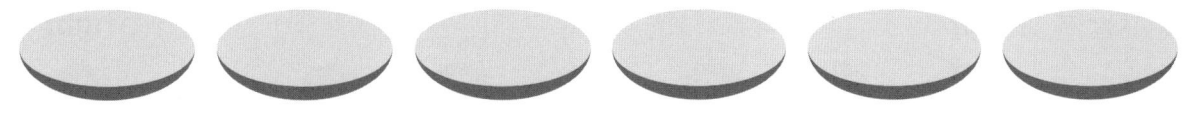

left　　　　　　　　　　　　　　　　　　　　　　　　　　　**right**

8 What is the position of the pencil?

The pencil is _____ from the top.

The pencil is _____ from the bottom.

9 Cross out the first three mice.

3rd

10 Match the 3rd key from the left to the last lock.

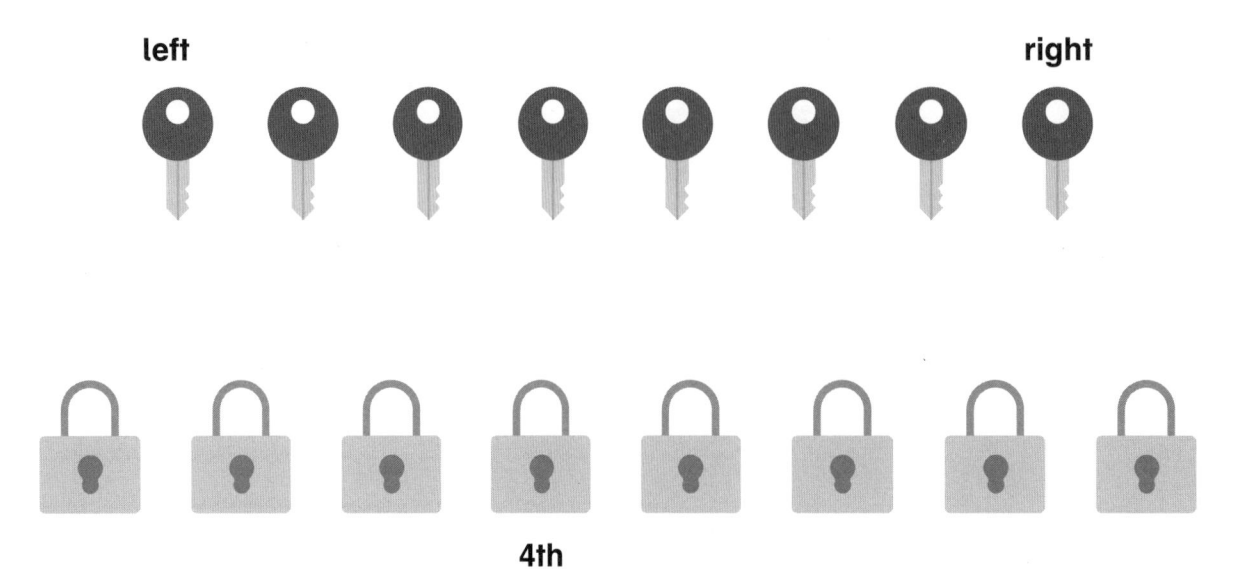

left   right

4th

126 Chapter 9 Test B

11 How many snails are to the left of the 2nd snail?

9th

There are _____ snails to the left of the 2nd snail.

12 Circle the child between the 5th and the 7th child.

1st

13 Sara is in line to buy tickets.
She is 11th from the end.
She is also first in line.
How many people are in line?

_____ people are in line.

14 Max lives in the 3rd house from the right.
Chapa lives in the 2nd house from the left.
How many houses are between them?

left **right**

There are _____ houses between them.

15 Matt and Grace are in line.
There are 2 people between Grace and Matt.
Grace is 1st in line and Matt is last in line.
How many people are in the line?

_____ people are in the line.

Name: _____

Date: _____

45 min   Score

60

Test A

Continual Assessment 2

Section A (2 points each)
Circle the correct option: **A**, **B**, **C**, or **D**.

1 How many tomatoes are there?

A eighteen

B nineteen

C sixteen

D twenty

2 7 is _____ less than 18.

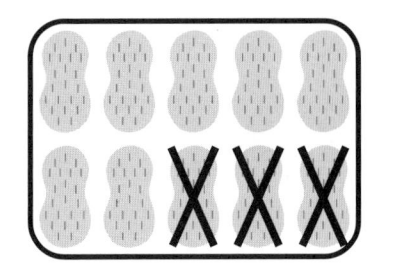

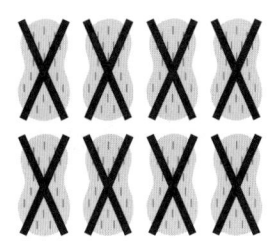

A 7

B 9

C 11

D 18

3 What are the missing numbers?
18, 17, _____ , 15, 14, _____

A 16 and 13

B 16 and 12

C 16 and 15

D 13 and 12

4 What is the missing number?

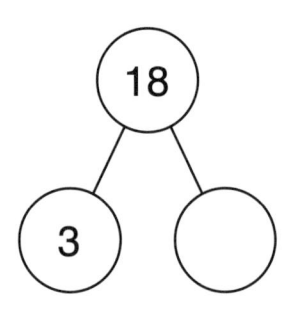

A 14

B 16

C 15

D 17

5 Which one will give the total number of chocolates in the two boxes?

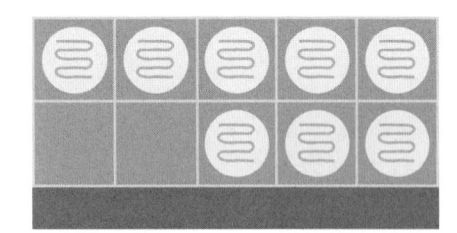

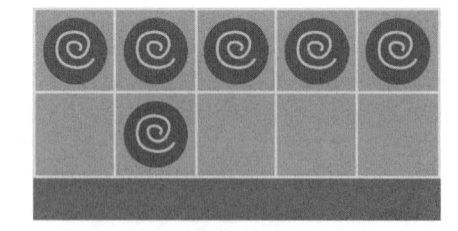

A 7 − 6

B 6 + 8

C 10 + 6

D 4 + 2

6 Which one will give the greatest answer?

    A 4 + 7   **B** 8 + 7

    C 6 + 5   **D** 7 + 9

7 A baker wants to put 12 donuts in a box.
There are already 8 donuts in the box.
How many more donuts does he need to add?

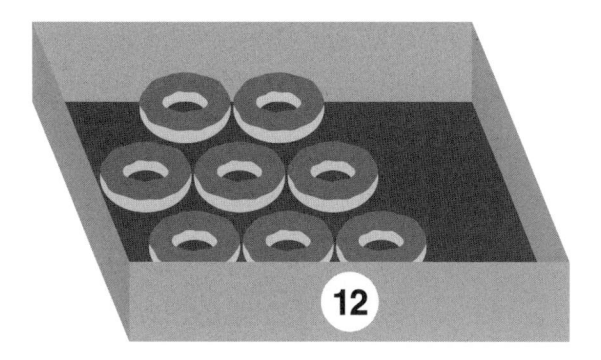

    A 6   **B** 8

    C 2   **D** 4

8 Which shape has 3 sides and 3 corners?

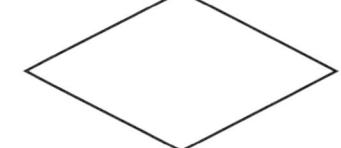

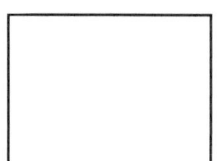

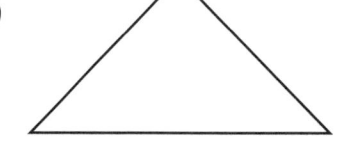

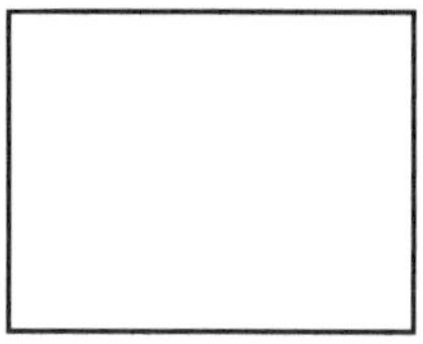

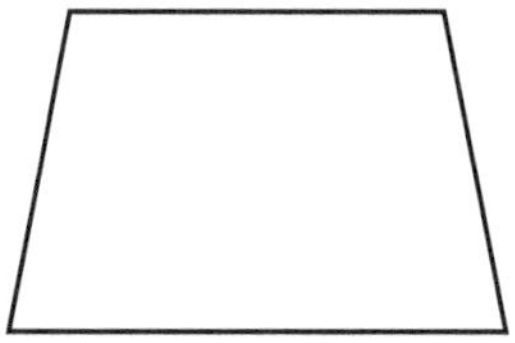

9 What shape is missing in the pattern?

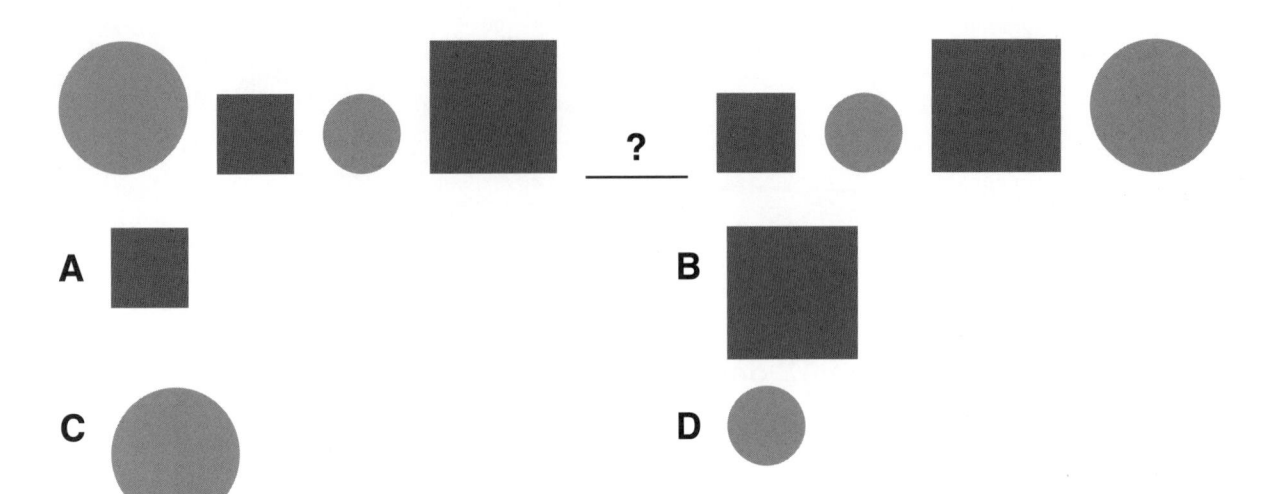

A ▪

B ◼

C ●

D •

10 The potato is _____ from the top.

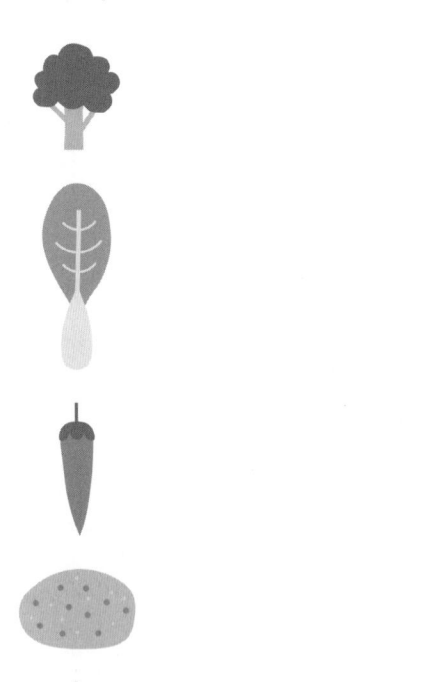

A 2nd

B 4th

C 3rd

D 1st

Continual Assessment 2 Test A

Section B (2 points each)

11 How many shells are there?
Circle the correct answer.

twelve

sixteen

fifteen

thirteen

12 Cross out the greatest number.
Circle the least number.

16   15   14
   12   19

Continual Assessment 2 Test A

13 Which set has fewer cupcakes?

Set A

Set B

Set _____ has fewer cupcakes than Set _____ .

14 Complete the number pattern.

| 10 | | 14 | | 18 | |

15 Complete the subtraction equation for this picture.

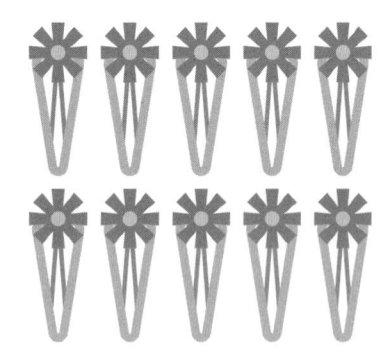

15 − ☐ = ☐

16 Write "+" or "−" in each circle.

5 ◯ 9 = 14

14 ◯ 5 = 9

17 Draw what comes next in the pattern.

18 Circle the 9th carrot.

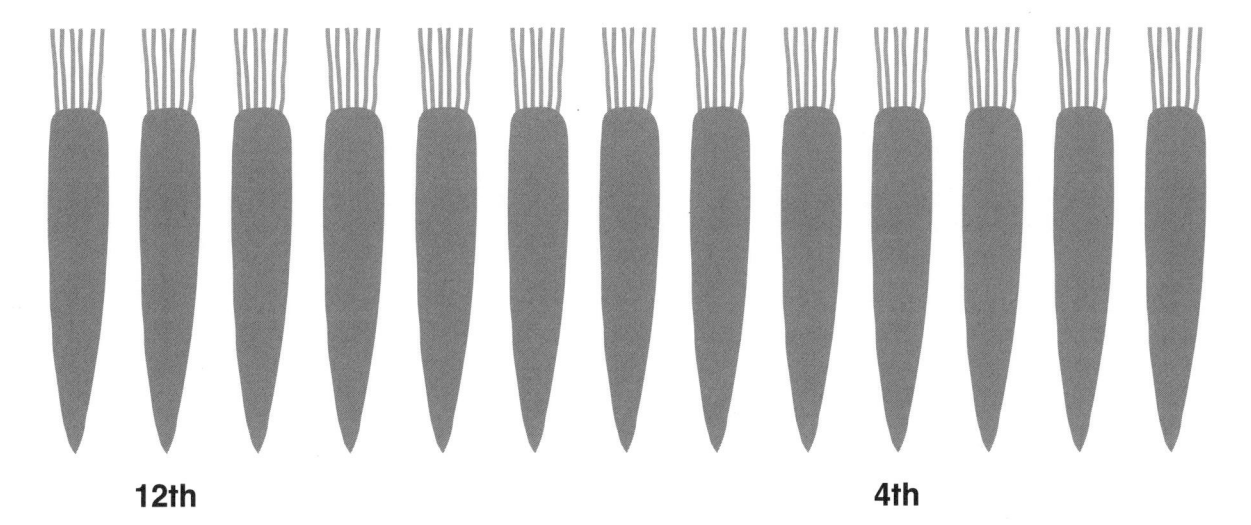

12th   **4th**

19 How many sides and corners does this figure have?

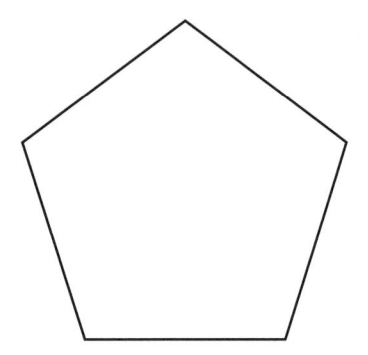

_____ sides

_____ corners

20 Complete the equation for this picture.

☐ − ☐ = 6

Section C (4 points each)

21 Siti has 15 apples.
9 of them are in a bag.
How many apples are not in the bag?

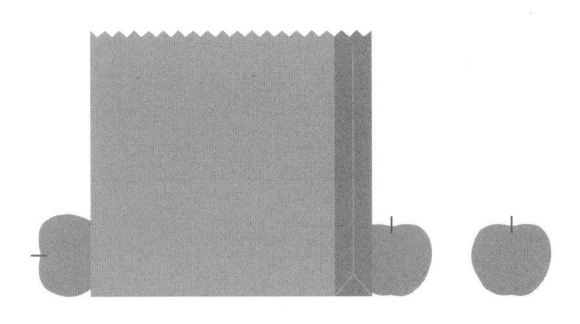

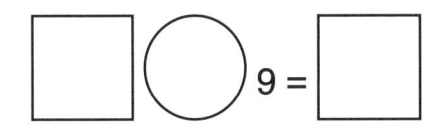

☐ ◯ 9 = ☐

_____ apples are not in the bag.

22 5 ladybugs are on top of a leaf.
7 ladybugs are hiding under the leaf.
How many ladybugs are there in all?

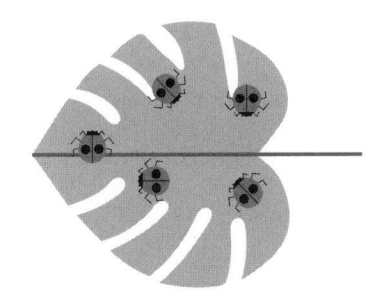

☐ ◯ ☐ = ☐

There are _____ ladybugs.

23 A farmer has 9 big watermelons in a basket.
She has 6 small watermelons in a box.
How many watermelons are there altogether?

☐ ◯ ☐ = ☐

There are _____ watermelons altogether.

24 There are 12 children in a room.
7 of them are girls.
How many boys are there in the room?

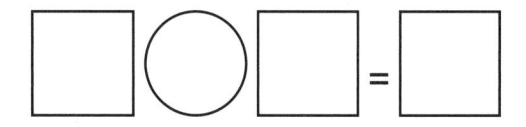

☐ ◯ ☐ = ☐

There are _____ boys in the room.

25 There are 8 people in a line.
Jacob is 4th from the front of the line.
What position is he from the back of the line?

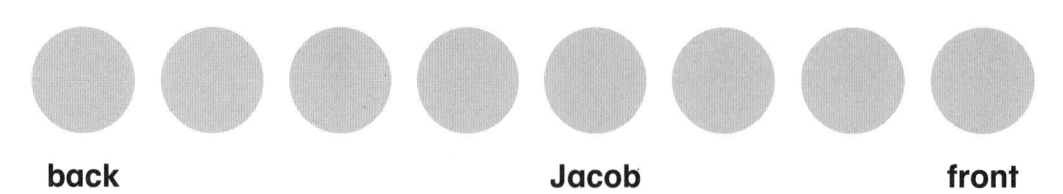

● ● ● ● ● ● ● ●

back   Jacob   front

Jacob is _____ from the back of the line.

Name: _____

Date: _____

45 min   Score

60

Test B

Continual Assessment 2

Section A (2 points each)
Circle the correct option: **A**, **B**, **C**, or **D**.

1 Which number is greater than 11 and less than 13?

  A ten

  B twenty

  C fourteen

  D twelve

2 After 1 dot is added to one of the cards, there will be _____ dots altogether.

  A 19

  B 16

  C 18

  D 10

3 20 − ☐ = 3 + 9

A 8

B 3

C 12

D 9

4 Which equation is false for this number bond?

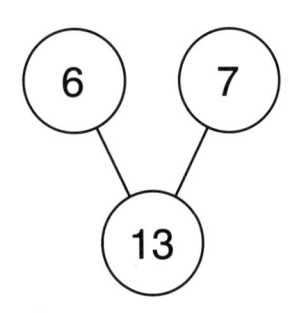

6 — 7
 \ /
  13

A 6 + 7 = 13

B 13 + 7 = 20

C 13 − 6 = 7

D 7 + 6 = 13

5 What numbers are missing in this number pattern?

| 9 | ☐ | 13 | 15 | ☐ | 19 |

A 12 and 16

B 11 and 18

C 11 and 17

D 18 and 20

6 Which 2 numbers together make 12?

A 7 and 5

B 6 and 7

C 8 and 3

D 9 and 9

7 Which equation is true?

A 15 − 6 = 13 − 7

B 12 − 4 = 15 − 8

C 13 − 4 = 15 − 6

D 11 − 7 = 17 − 9

8 Which shape comes next in this pattern?

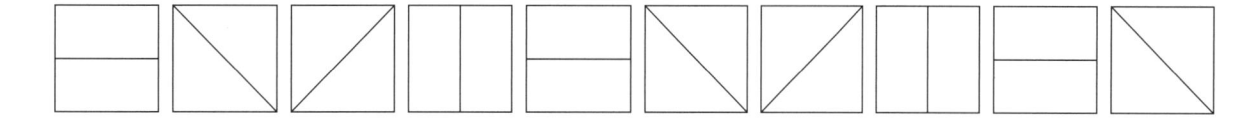

A

B

C

D

9 How many mice are between the 5th and the 11th mice?

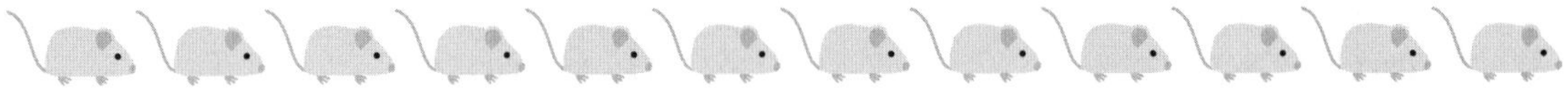

1st

A 11

B 7

C 10

D 5

10 Which shape has more than 4 corners?

A

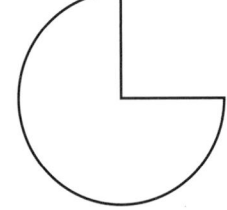

B

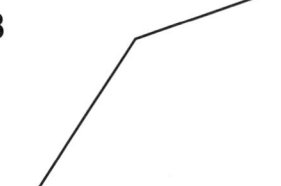

C

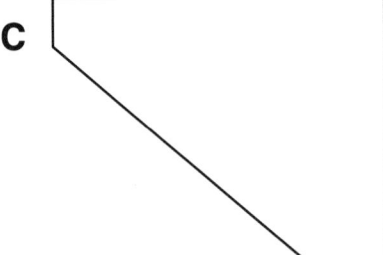

D

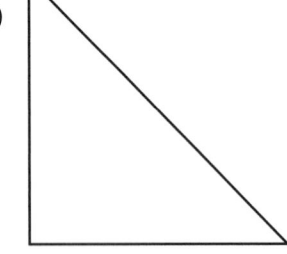

Section B (2 points each)

11 Write a number from the box to make each sentence true.

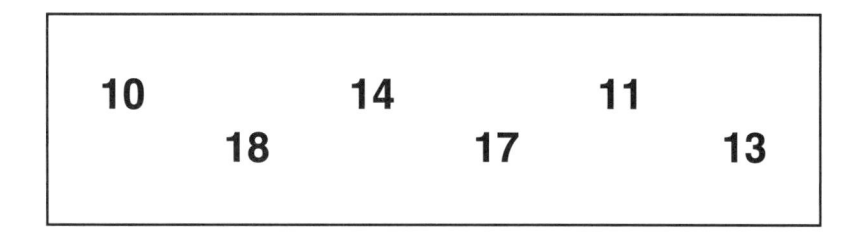

10	14	11
18	17	13

☐ is the least number.

☐ is the greatest number.

12 Circle the number that is 3 more than 17.

fourteen   fifteen   nineteen   thirteen   twenty

13 Write the numbers in order from greatest to least.

10	19	20
16	0	

☐ ☐ ☐ ☐ ☐

Continual Assessment 2 Test B

14 Cross out the boxes that have an answer of 5.

| 13 – 8 | 14 – 7 | 12 – 9 | 11 – 6 |

15 Circle the shape that comes next in the pattern.

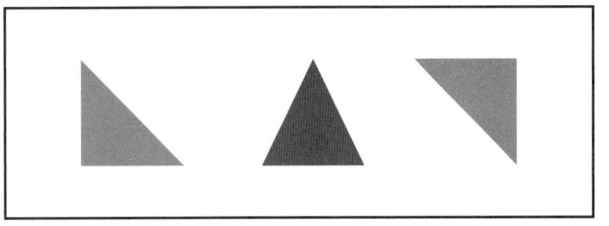

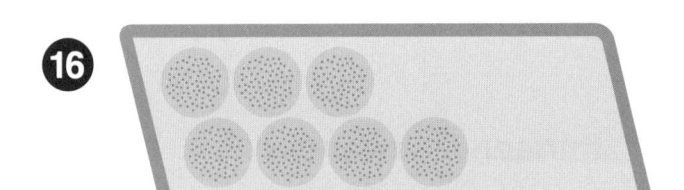

16

Tray A

Tray B

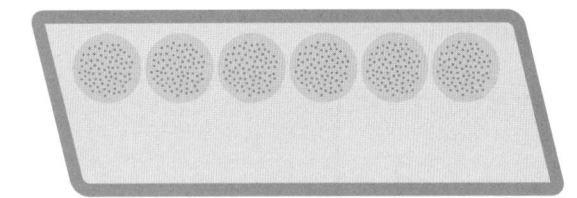

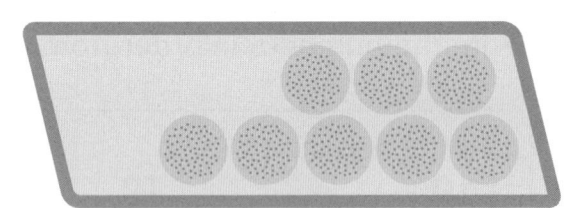

Tray C

Tray D

Tray _____ has the most cookies.

Tray _____ and Tray _____ have fewer cookies than Tray D.

Continual Assessment 2 Test B

17 Draw a line to cut this figure into a square and a triangle.

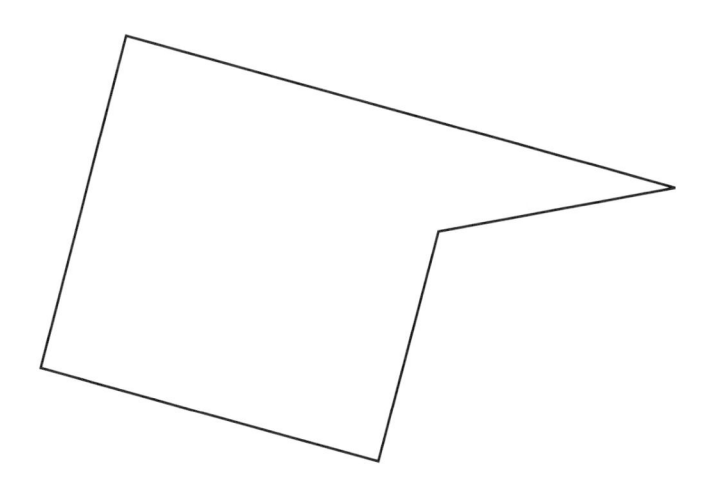

18 How many squirrels are hidden behind the rock?

14th **2nd**

_____ squirrels are hidden.

19 Cross out the shape that does not belong.

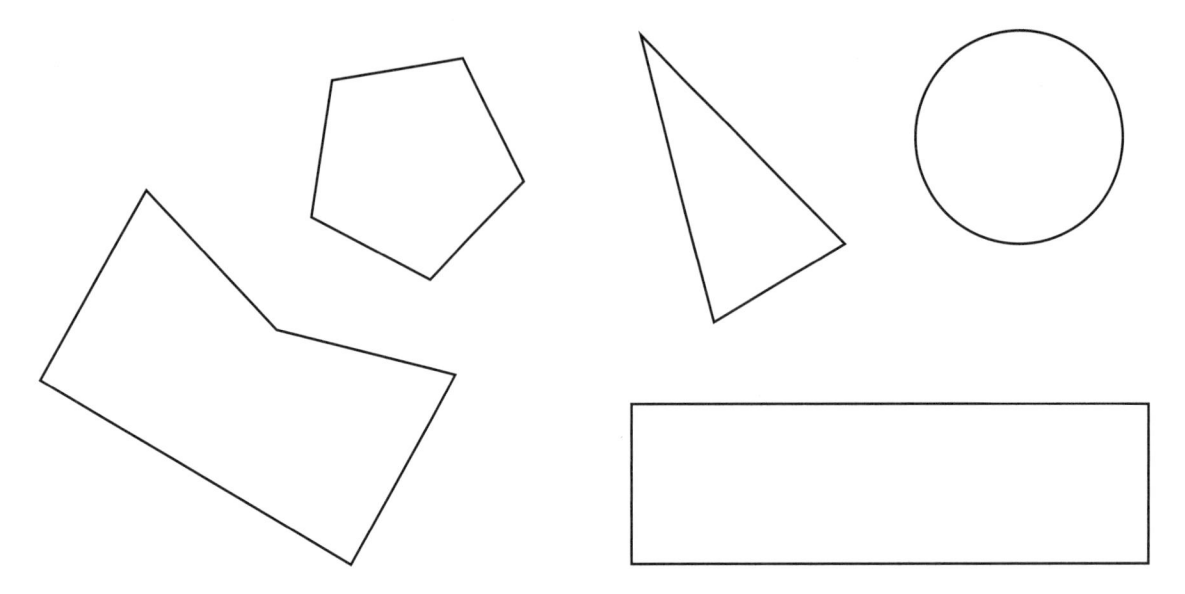

20 Mei wins two prizes at the state fair.
She wins the prize in the middle of the row.
She also wins the prize 3rd from the left.
Circle the 2 prizes that she wins.

left   right

Section C (4 points each)

21 How many circles, triangles, and rectangles are there in the picture?

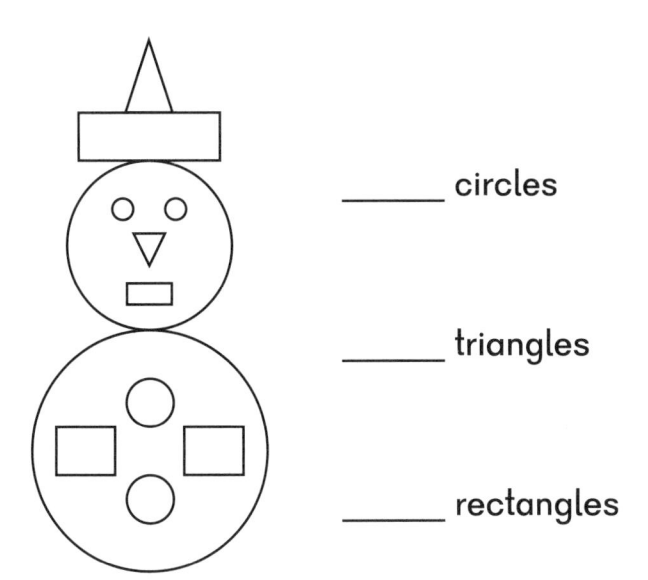

_____ circles

_____ triangles

_____ rectangles

22 There are 12 balloons.
Some balloons fly away.
8 balloons are left.
How many balloons fly away?

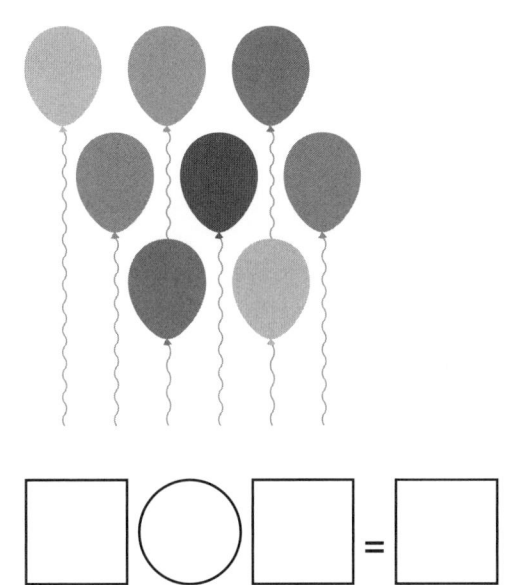

☐ ○ ☐ = ☐

_____ balloons fly away.

23 There are 11 flowers in a vase.
5 of the flowers are red.
The rest are white flowers.
How many white flowers are in the vase?

☐ ◯ ☐ = ☐

There are _____ white flowers in the vase.

24 There are 4 people standing in line.
9 more people join the line.
How many people are in line now?

☐ ◯ ☐ = ☐

There are _____ people in line now.

25 Lily is 8th from the front of a line.
She is 4th from the back of the line.
How many people are waiting in line?

There are _____ people waiting in line.

Answer Key

Test A

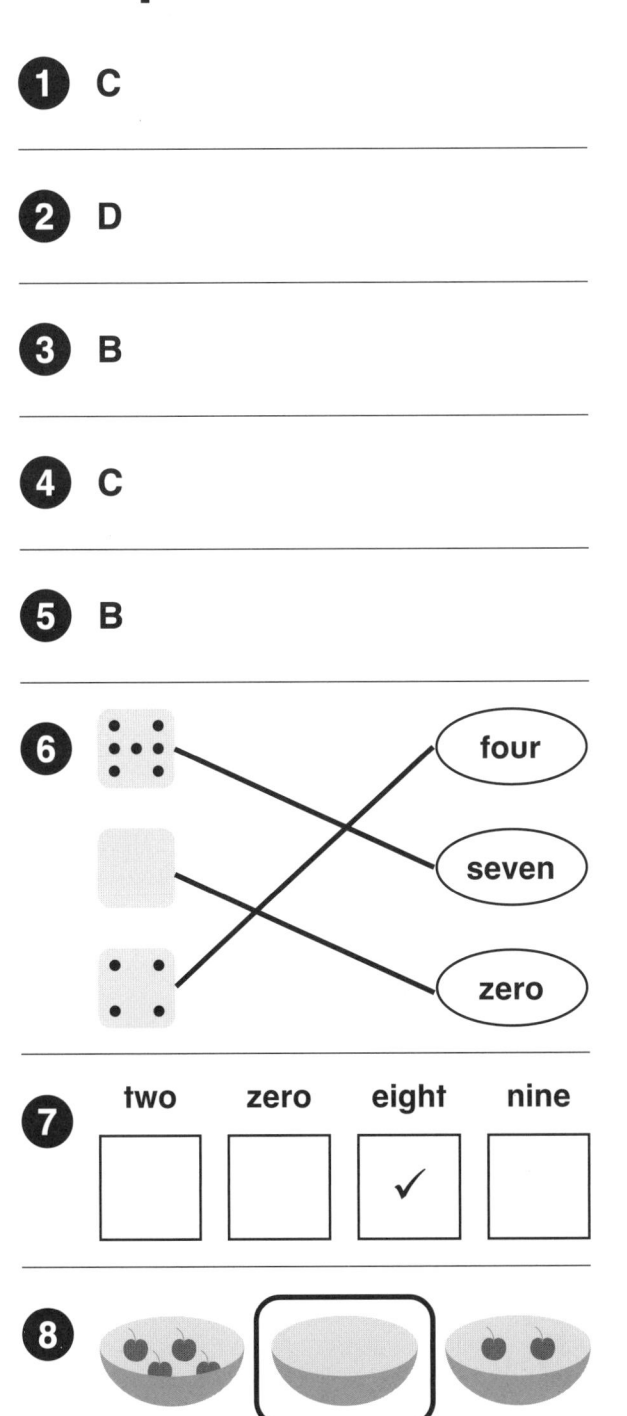

Chapter 1 Numbers to 10

1. C

2. D

3. B

4. C

5. B

6. 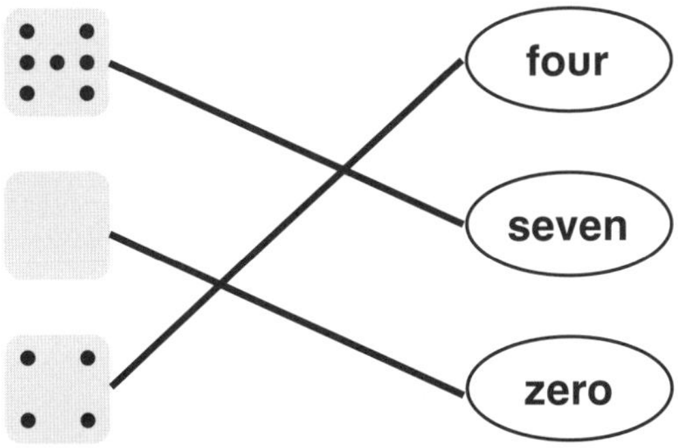

7.

two	zero	eight	nine
		✓	

8. 

9. three 3 five 5

10. (7) 5 0 4

11. 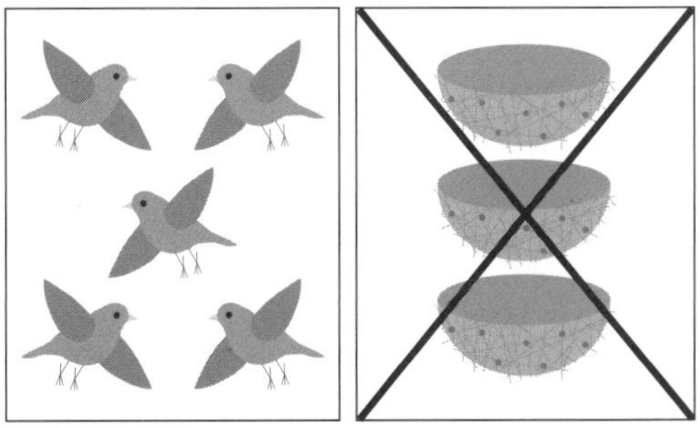

12. 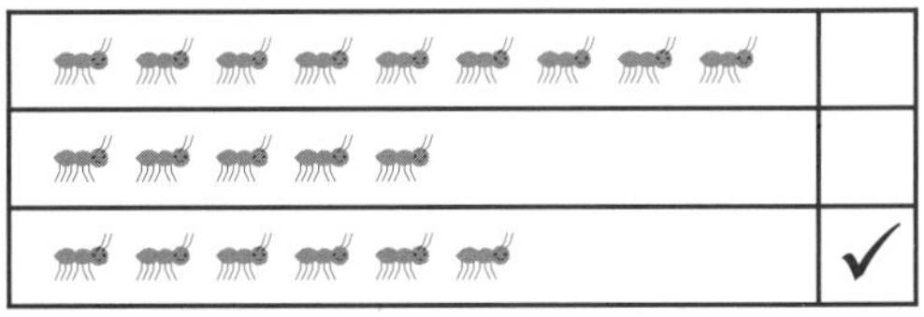

13. (three) five ten six

14. 9

15. 2 3 **4** 5 6 **7**

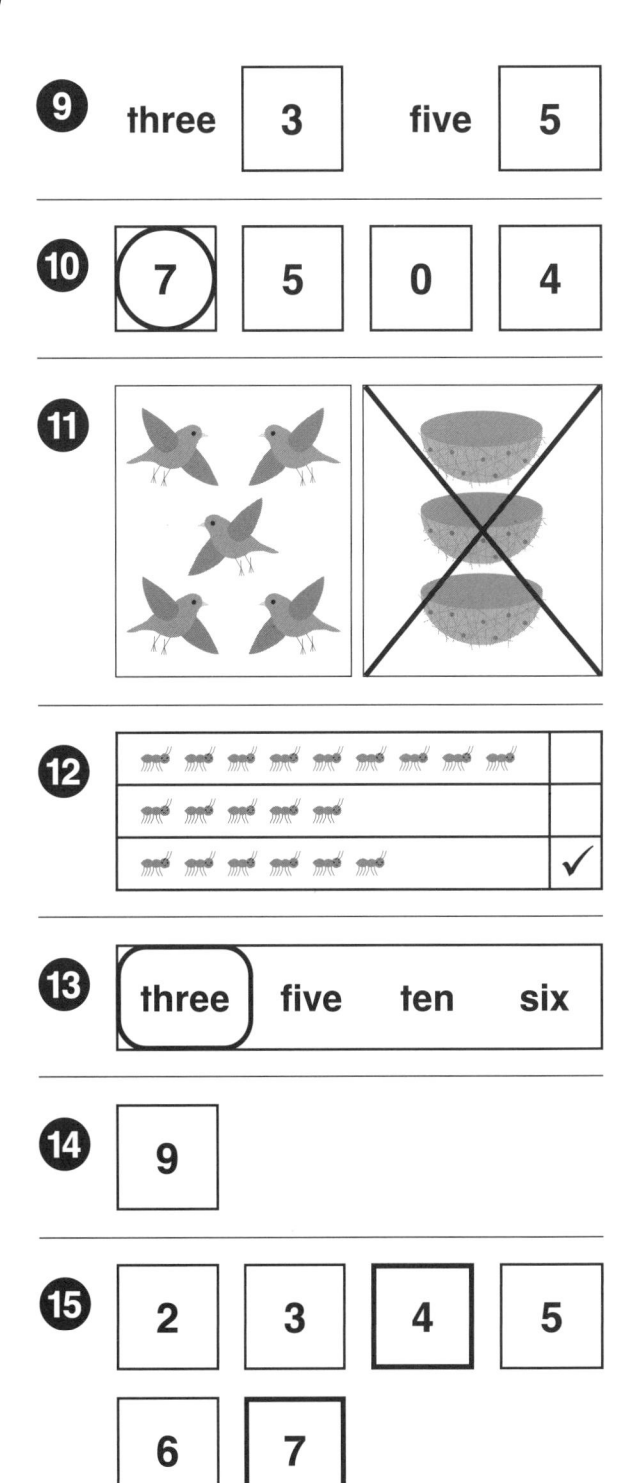

Test B

Chapter 1 Numbers to 10

1 B

2 B

3 D

4 A

5 B

6 zero — 0

ten — 10

7 six

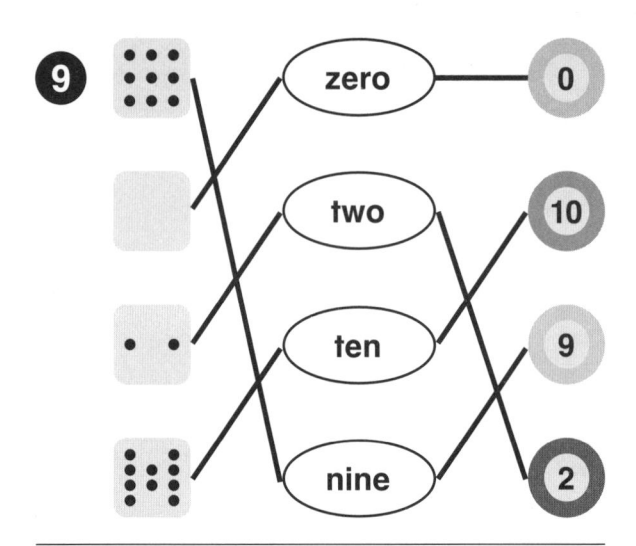

8 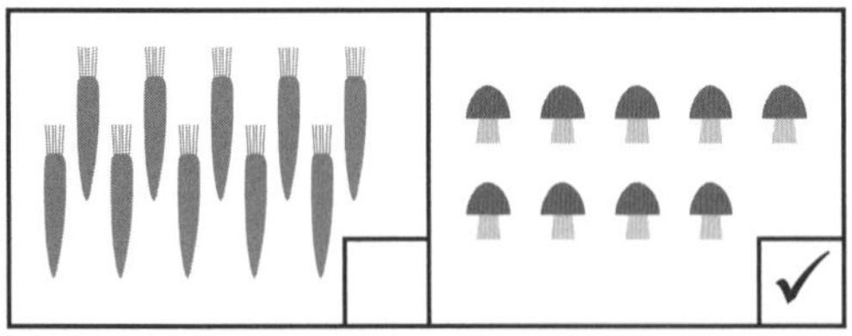 (mushrooms checked ✓)

9 (matching: dice of 9 → nine → 9; two → 2; ten → 10; zero → 0)

10 6, 4 (boxed)

11 (strawberries)

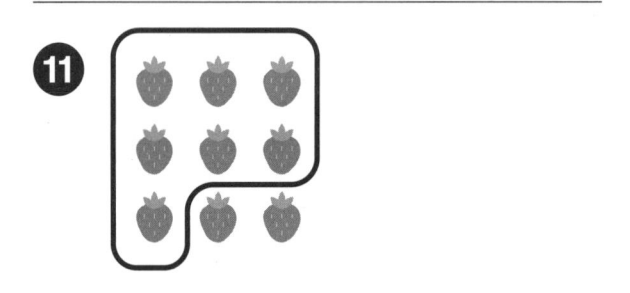

12 7, 5, 0, ⊠, 8

13 9

14 0

15 5, 6, and 7

Test A

Chapter 2 Number Bonds

1 B

2 D

3 B

4 C

5 A

6 4 and **2** make 6.

7 3 and **4** make 7.

8 5 is 1 and **4**.

9 4, 3, 1

10 3, 9, 6

11 (7 apples split 5 and 2)

7 — 5, 2

12 **5** and **5** make 10.

13 6 and 4 circled (6, 1, 3, 2, 4) — actually 6 and 2 circled

14 6 — 2, 4

15 10 — 0, 10

Test B

Chapter 2 Number Bonds

1 A

2 C

3 D

4 A

5 C

6 6 is [4] and 2.

7 4 and 4 make [8].

8

6	4	3
1	5	6
1	9	8

9 (7) 2 6 5 (3)

10 6 — 3, 3

11 0 — 10, 10

12 1 — 1, 0

13 7 — 2, 5 — 1, 4

14 2, 5 — 7 — 10, 3

15 6 — 3, 3

Test A

Chapter 3 Addition

1 B

2 B

3 D

4 B

5 B

6 3 + 5 = 8

7 6 is 2 more than 4.

8 5, 5, 0

0 + 5 = 5

9 6 + 3 = 9

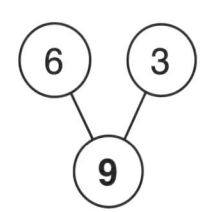

6, 3, 9

10 4 + 2 = 6

11 2 + 3 = 5

12 6 + 4 = 10

13 2 + 7 = 9

There are 9 berries on the two plates.

14 5 + 1 = 6

There are 6 bats in all.

15 2 + 7 = 9

There are 9 watermelons altogether.

Test B

Chapter 3 Addition

1 C

2 C

3 B

4 B

5 C

6 3 + [2] = [5]

7 [4] + [4] = [8]

8 7 + [3] = [10]

9 [1] + 6 = [7]

10 2 more than 5 is [7].

11 8 and 1 make [9].

12 6 + 2 = [1] + 7

13 9 + 1 = 10
1 + 9 = 10

14 [0] + [3] = [3]

There are 3 cookies in the jar altogether.

15 [6] (+) [4] = [10]

Dion has 10 baseball cards in all.

Test A

Chapter 4 Subtraction

1 B

2 D

3 A

4 B

5 C

6 8 − 5 = **3**

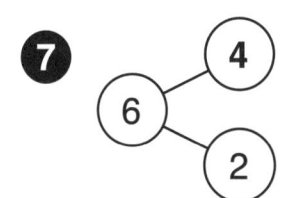

7 6 → 4, 2

6 − **2** = 4

8 **10** − 9 = **1**

9 7 − **5** = **2**

7 − **2** = **5**

10 5 **+** 3 = 8

3 **+** 5 = 8

8 **−** 5 = 3

8 **−** 3 = 5

11 6 − 2 = 4

12 5 − **5** = **0**

0 balloons are left.

13 8 − 2 = 6

6 of her hair bands are not in her pocket.

14 9 **−** 5 = 4

There are **4** boys in the room.

15 10 **−** 2 = 8

There are **8** children left on the bus.

Test B

Chapter 4 Subtraction

1 C

2 D

3 C

4 A

5 C

6 9 − 7 = 2

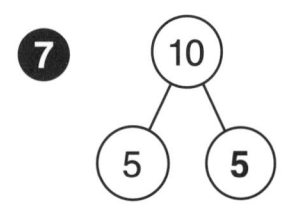

7
- 10 = 5 and 5
- 10 − 5 = 5

8 8 − 3 = 5

5 eggs have not hatched.

9 5 − 2 = 3

5 − 3 = 2

10 3 + 3 = 6

6 − 3 = 3

11 7 + 3 = 10

3 + 7 = 10

10 − 3 = 7

10 − 7 = 3

12 8 − 3 = 5

There are 5 children not wearing hats.

13 9 − 5 = 4

4 children are still at the party.

14 10 − 4 = 6

Sofia has 6 baseball cards.

15 7 − 2 = 5

5 birds fly away.

Test A

Continual Assessment 1

1 B

2 C

3 A

4 D

5 B

6 D

7 B

8 D

9 C

10 B

11 (eight)　seven　ten　nine

12

13 (9)　5　6　8　(10)

14 2　3　4　5　　6　7

15 3　5　7　9

16 3, 4 — 7

17 10 — 1, 9

18 4, 4 — 8

19 $4 + 3 = 7$

20 $5 - 2 = 3$

Test A

Continual Assessment 1

21) 7 (+) [3] = [10]

There are <u>10</u> limes in all.

22) [4] (−) [1] = [3]

<u>3</u> eggs are not broken.

23) [9] (−) [6] = [3]

<u>3</u> apples fell off the tree.

24) [4] (+) [6] = [10]

Emma has <u>10</u> stickers.

25) [7] (−) [2] = [5]

There are <u>5</u> slices of pizza left.

Test B

Continual Assessment 1

1 D

2 C

3 B

4 A

5 D

6 A

7 D

8 B

9 C

10 A

11 10 9 8 7
       6 5

12 ✓ (top row checked)

13 0 5 6 8
       10

14 4 ⊗ 2 5

15 <u>9</u> is more than 8 and less than 10.

16
- 1 — 5
- 1 — 4
- 2 — 3
- 1

17
- 5 — 3 — 2
- 7 — 2, 5

Continual Assessment 1 Test B

159

Test B

Continual Assessment 1

18 | 2 8 ④ 6 ⑤ |

19 0 + [3] = [3]

[3] + [0] = [3]

20 [9] − 8 = [1]

[9] − [1] = [8]

21 [3] (+) [2] = [5]

There are 5 owls in all.

22 [6] (−) [1] = [5]

There are 5 shells still left in the bucket.

23 [3] (+) [5] = [8]

Kona has 8 pumpkins altogether.

24 [4] (+) [2] = [6]

There are 6 ducks in all.

25 [10] (−) [7] = [3]

Sam has 3 oranges.

Test A

Chapter 5 Numbers to 20

1 B

2 C

3 A

4 A

5 D

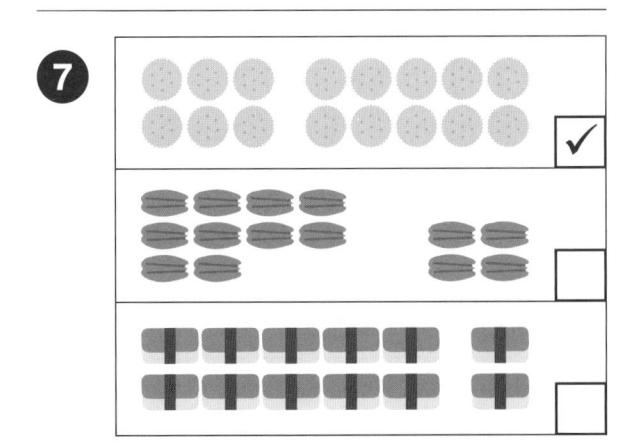

6 $\boxed{17} - 7 = \boxed{10}$

7 (top row ✓)

8 12, 16, ~~13~~, 15

9 (left box ✓)

10 11, 14, 17, 19

11 13, ⟨20⟩, 14, 12, ⟨16⟩

12 18, 17, 16, 15, 14, 13

13 20

14 $\boxed{10} + \boxed{3} = \boxed{13}$

Ella has 13 beads.

15 $\boxed{15} - \boxed{4} = \boxed{11}$

11 birds are still in the tree.

Test B

Chapter 5 Numbers to 20

1 D

2 B

3 C

4 B

5 C

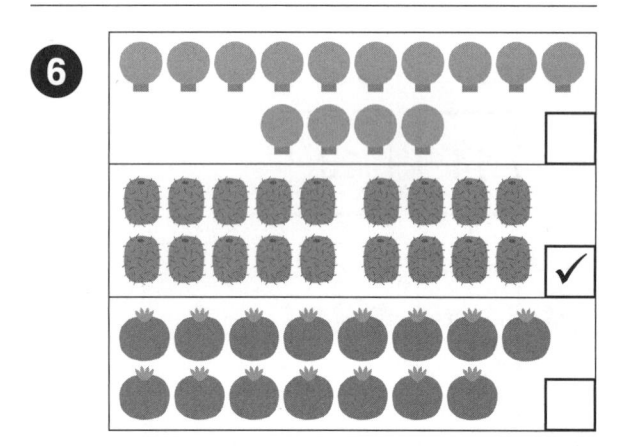

6 (second row checked)

7 5 less than 18 is 13.

8 19 / 10   9

9 17 − 3 = 14

10 11 + 4 = 19 − 4

11 6   8   [10]   12   [14]   [16]

12 10   12   17   18   20

13 11   (15)   (16)   20   19

14 15 − 5 = 10

Chapa has 10 crayons.

15 13 (+) 4 = 17

He has 17 red apples.

Test A

Chapter 6 Addition to 20

1 B

2 C

3 C

4 A

5 C

6 (11 and 0) circled; (4 and 7) circled

7 12

8 7 (13) 16 10 — 13 circled

9 8 + 8 = 16

10 4 + 9 = 10 + 3; branches 3 and 1

11 (6 + 9) circled; (7 + 8) circled

12 8 + 4 = 12

13 6 + 5 = 11

14 2 + 9 = 11

There are 11 candies in all.

15 6 + 6 = 12

There are 12 animals altogether.

Test B

Chapter 6 Addition to 20

1 C

2 D

3 A

4 D

5 C

6 | 7 + 5 | 9 + 6 | ~~2 + 9~~ |

7 | 7 + 7 | (6 + 9) | 8 + 4 |

8 4 + 6 = 1 + [9]

9 8 + 4 = 7 + [5]

10 4 more than 13 is the same as 5 more than 12.

11
13 + 3 = 13	☐
6 + 9 = 15	✓
18 = 8 + 8	☐

12
| 7 + 4 | (8 + 6) |
| 0 + 13 | 5 + 6 |

13 [4] + [12] (=) [16]

There are 16 fish altogether.

14 [7] (+) [8] (=) [15]

She buys 15 oranges and apples in all.

15 [4] (+) [9] (=) [13]

Alex has 13 stickers altogether.

Test A

Chapter 7 Subtraction Within 20

1 A

2 B

3 D

4 A

5 A

6 13 − 5 = **8**

7 16 − 9 = **7**

8 13 − **9** = 4

9 9   14   16   **(6)**

10 6 less than 12 is <u>6</u>.

11 | 11 − 4 | | 16 − 5 |

| 18 − 8 | **(17 − 9)**

12 **13** − 4 = **9**

13 **14** − **6** = **8**

There are <u>8</u> slices of watermelon left.

14 **15** **(−)** **7** = **8**

<u>8</u> crabs are left.

15 **12** **(−)** **9** **(=)** **3**

<u>3</u> children get off the bus.

Test B

Chapter 7 Subtraction Within 20

1 B

2 A

3 D

4 A

5 A

6 twelve | fourteen | **fifteen** (circled) | seventeen

7 **14 – 6** (circled) | 17 – 8 | **15 – 7** (circled) | 14 – 9

8 11 – 9 = 8 (crossed out) | 16 – 9 = 7

9 $\boxed{13} - 6 = 7$

10 $14 - 7 = 12 - \boxed{5}$

11 $15 - \boxed{6} = 6 + 3$

12 $\boxed{11} - \boxed{2} = \boxed{9}$

She has 9 apples left.

13 $12 - \boxed{5} = \boxed{7}$

She needs to add 7 fish.

14 $14 - 3 = 11$

Dion has 11 stickers left.

15 $15 - 7 = 8$

Jacob needs to remove 8 books.

Test A

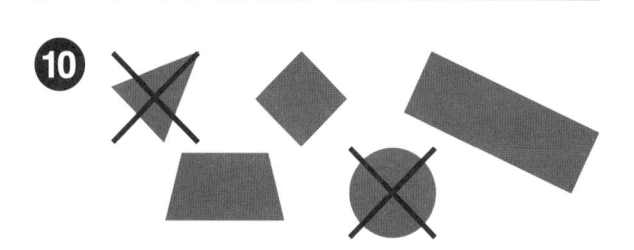

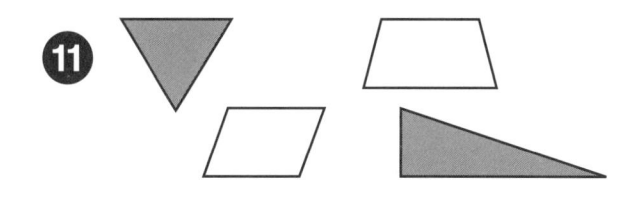

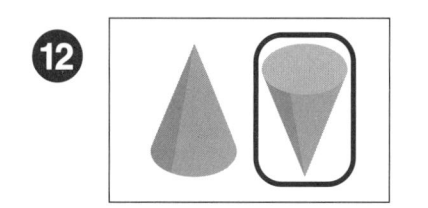

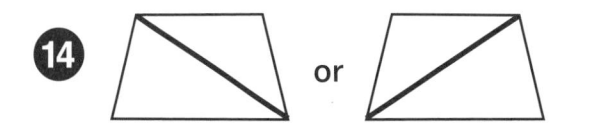

Chapter 8 Shapes

1 C

2 C

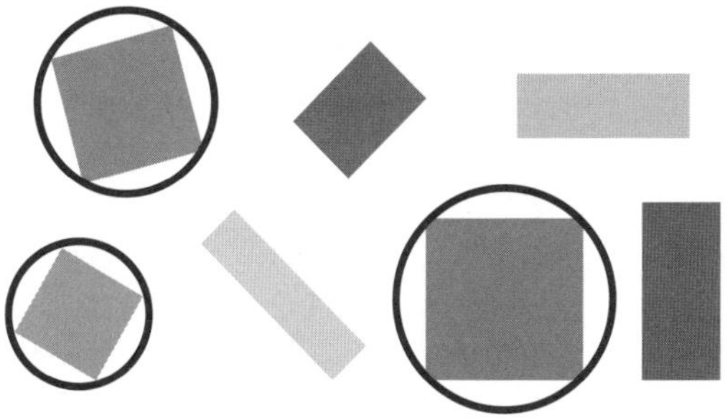

3 D

4 C

5 D

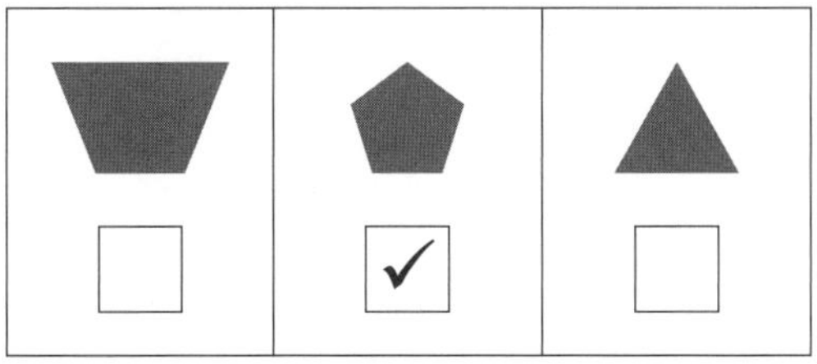

9 This picture has <u>6</u> circles.

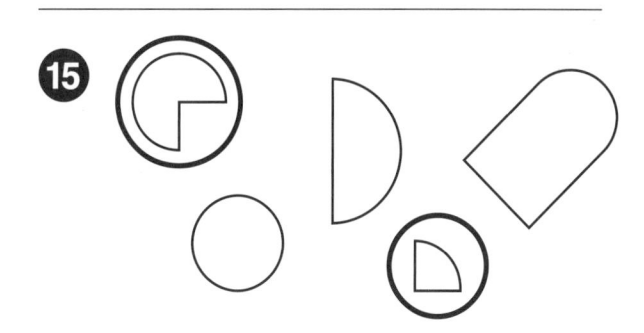

Test B

Chapter 8 Shapes

1 B

2 A

3 A

4 D

5 C

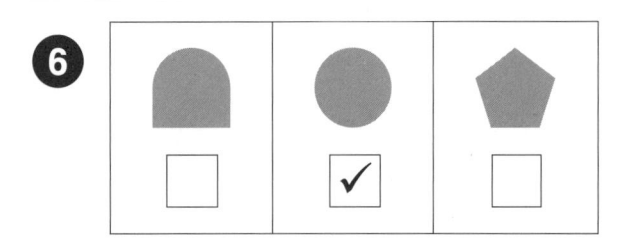

6 (checkmark on middle shape - circle)

7 (rectangle and triangle)

8 <u>3</u> rectangles
<u>5</u> triangles

9 A rectangle has <u>4</u> corners.

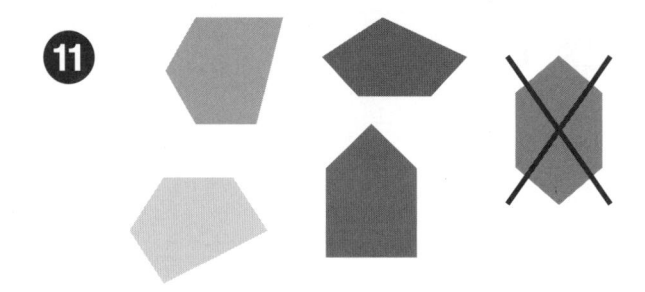

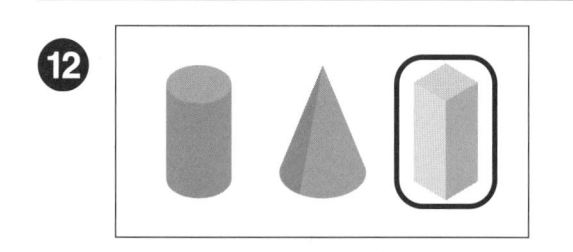

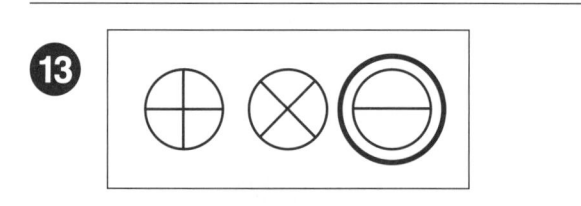

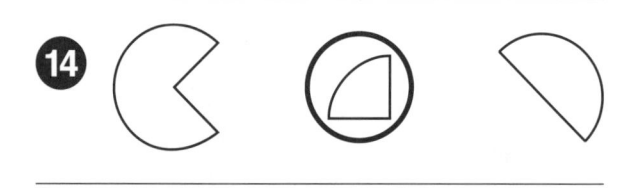

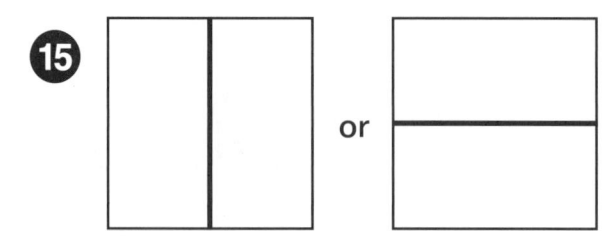

10 <u>5</u> corners
<u>5</u> sides

11 (X on third shape)

12 (circle around third shape - rectangular prism)

13 (circle around third shape)

14

15

Test A

Chapter 9 Ordinal Numbers

1 C

2 C

3 B

4 D

5 C

6 (2nd dog circled)

7 (last 5 chicks crossed out)

8 (apple in 4th box)

9 (2nd ball crossed out)

10 (line from bird to 6th nest)

11 There are 2 cars between the 2nd car and the 5th car.

12 (6th robot crossed out)

13 (2nd cube crossed out)

14 There are 5 people in the line.

15 There are 9 people in the line.

Test B

Chapter 9 Ordinal Numbers

1 A

2 B

3 C

4 A

5 B

6 3 animals are behind the giraffe.

7

8 The pencil is 3rd from the top.
The pencil is 2nd from the bottom.

9

10

11 There are 7 snails to the left of the 2nd snail.

12

13 11 people are in line.

14 There are 4 houses between them.

15 4 people are in the line.

Test A

Continual Assessment 2

1 B

2 C

3 A

4 C

5 B

6 D

7 D

8 D

9 C

10 B

11 sixteen

12 16   (12)   15   ~~13~~   14

13 Set <u>A</u> has fewer cupcakes than Set <u>B</u>.

14 10 12 14 16 18 20

(12, 16, 18, 20 bolded)

15 $15 - \boxed{2} = \boxed{13}$

16 $5 \; \boxed{+} \; 9 = 14$

$14 \; \boxed{-} \; 5 = 9$

17

18

19 <u>5</u> sides

<u>5</u> corners

Test A

Continual Assessment 2

20) $\boxed{12} - \boxed{6} = 6$

21) $\boxed{15} \;\bigcirc\!{-}\; 9 = \boxed{6}$

<u>6</u> apples are not in the bag.

22) $\boxed{5} \;\bigcirc\!{+}\; \boxed{7} = \boxed{12}$

There are <u>12</u> ladybugs.

23) $\boxed{9} \;\bigcirc\!{+}\; \boxed{6} = \boxed{15}$

There are <u>15</u> watermelons altogether.

24) $\boxed{12} \;\bigcirc\!{-}\; \boxed{7} = \boxed{5}$

There are <u>5</u> boys in the room.

25) Jacob is <u>5th</u> from the back of the line.

Test B

Continual Assessment 2

1 D

2 C

3 A

4 B

5 C

6 A

7 C

8 B

9 D

10 C

11 10 is the least number.

18 is the greatest number.

12 fourteen   fifteen   nineteen   thirteen   (twenty)

13 20 19 16 10

0

14 ~~13 − 8~~   14 − 7   12 − 9   ~~11 − 6~~

15

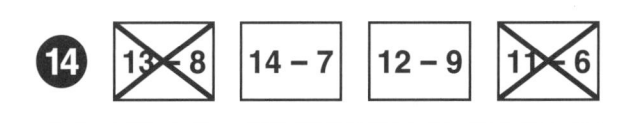

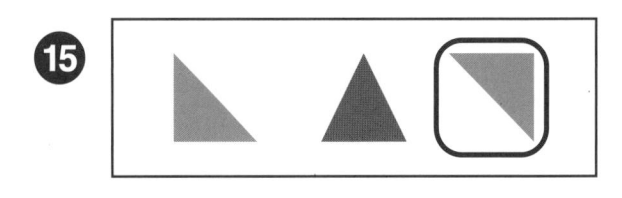

16 Tray B has the most cookies.

Tray A and Tray C have fewer cookies than Tray D.

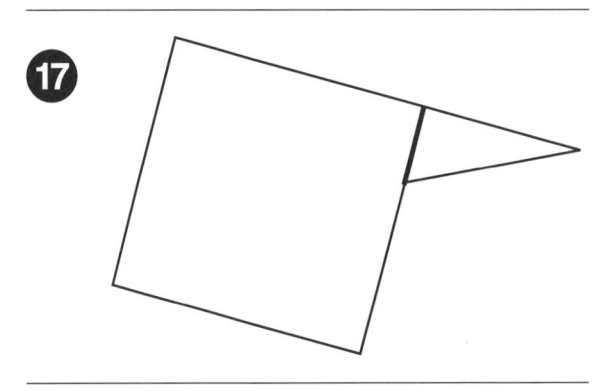

17

18 6 squirrels are hidden.

Test B

Continual Assessment 2

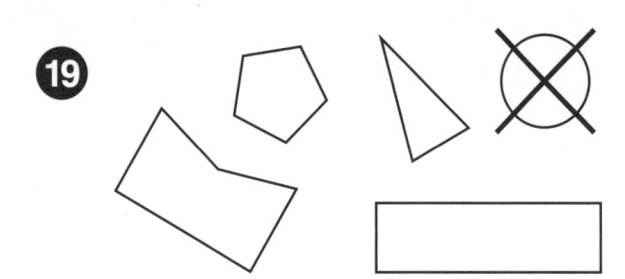

19

20

21 6 circles
2 triangles
4 rectangles

22 | 12 | (−) | 8 | = | 4 |

4 balloons fly away.

23 | 11 | (−) | 5 | = | 6 |

There are 6 white flowers in the vase.

24 | 4 | (+) | 9 | = | 13 |

There are 13 people in line now.

25 There are 11 people waiting in line.